NONLINEAR
DYNAMICAL SYSTEMS

NONLINEAR DYNAMICAL SYSTEMS

Feedforward Neural Network Perspectives

Irwin W. Sandberg

James T. Lo

Craig L. Fancourt

Jose C. Principe

Shigeru Katagiri

Simon Haykin

A WILEY-INTERSCIENCE PUBLICATION

JOHN WILEY & SONS, Inc.

New York / Chichester / Weinheim / Brisbane / Singapore / Toronto

Library of Congress Cataloging-in-Publication Data:

Nonlinear dynamical systems: feedforward neural network perspectives / Irwin Sandberg . . . [et al.].
    p. cm.
  Includes bibliographical references and index.
  ISBN 0-471-34911-9 (cloth: alk. paper)
  1. Neural networks (Computer science) 2. Dynamics. I. Sandberg, Irwin.
QA76.87.N56 2001
006.3′2—dc21

                                      00–043384

10 9 8 7 6 5 4 3 2 1

CONTENTS

PREFACE

Feedforward neural networks have established themselves as an important part of the rapidly expanding field of artificial neural networks. This book *Nonlinear Dynamical Systems: Feedforward Neural Network Perspectives* addresses fundamental aspects and practical applications of the subject. To the best of the authors' knowledge, this is the first book to be published in this area.

Chapter 1 provides an introductory treatment of the different aspects of feedforward neural networks, thereby setting the stage for more detailed treatment of the subject matter in the succeeding four chapters.

Chapter 2 is concerned with classification problems and with the related problem of approximating dynamic nonlinear input–output maps. Attention is focused on the properties of nonlinear structures that have the form of a dynamic preprocessing stage followed by a memoryless nonlinear section. It is shown that an important type of classification problem can be solved using certain simple network structures involving linear functionals and memoryless nonlinear elements. The chapter addresses various aspects of the problem of approximating nonlinear input–output maps. One main result given is a theorem showing that if the maps to be approximated satisfy a certain "myopicity" condition, which is very often met, then they can be uniformly approximated arbitrarily well using a structure consisting of a linear preprocessing stage followed by a memoryless nonlinear network. Noncausal as well as causal maps are considered. (Approximations for noncausal maps are of interest in connection with image processing.) In the course of the study on which the material of this chapter is based, some interesting unexpected side issues arose. One such issue is discussed in an appendix where, in connection with the study of myopic maps, attention is focused on—and a correction is given of—a longstanding oversight concerning the cornerstone of digital signal processing. The chapter makes use of concepts drawn from the areas of real analysis and functional analysis.

Chapter 3 relates to one of the major research areas in the past 15 years, which pertains to the development of robust controllers and filters. Robust controllers and filters are intended to avert disastrous results or mitigate worst-case performances. The H-infinity norms, minimax errors, and risk-sensitive functionals are the main criteria used to induce robust performance and are proven to lead to the same robust controllers and filters for linear systems. Extending such results to nonlinear systems by the conventional analytic approach also has been a topic of extensive research. Dynamic programming equations characterizing the robust controllers and filters have been obtained. However, these equations are difficult, if not impossible, to use to derive practically useful results. In Chapter 3, the capabilities of neural networks to approximate functions and dynamic systems to any accuracy with respect to risk-sensitive error specifically are discussed. It is shown that under mild conditions, a function can be approximated, to any desired degree of accuracy with respect to a general risk-sensitive criterion, by a multilayer perceptron or a radial-basis function network. It is also shown that under relatively mild regularity conditions, dynamic systems can be approximated (or identified) by neural networks to any desired degree of accuracy with respect to general risk-sensitive criteria in both the series-parallel and parallel formulations. These capabilities of neural networks for universal risk-sensitive approximations of functions and dynamic systems qualify neural networks as powerful vehicles in a synthetic approach to robust processing (e.g., signal processing, communication, and control).

Chapter 4 discusses the practical issue of segmenting a time series. In this context, we note that many of the methods for detecting abrupt statistical changes in time series evolved in the field of statistical quality control in the 1950s. In fact, much of the nomenclature still reflects this origin. For example, in the statistics and control literature, time-series classification is often referred to as "isolation," which is short for "fault isolation." At that time, the emphasis was on the simpler problem of detecting changes in the moment(s) of an independent process, such as the dimension of a part coming off an assembly line. As digital signal processing advanced in the 1970s, the change detection methods were extended to include processes with memory. However, such methods generally still utilized linear models. In the 1990s, with the advent of powerful and general nonlinear modeling techniques, such as neural networks, new multiple-model algorithms appeared for modeling nonlinear but piecewise stationary time series. These algorithms were even able to model switching chaotic signals but seemingly had no connection with the prior work in quality control. Thus, the goal in Chapter 4 is twofold. From one side, we reexamine the classical methods for modeling piecewise stationary signals with an eye toward integration with new nonlinear models. We then push these algorithms into the realm of chaotic signals and examine whether they still function as before and why. From the other side, we put many of the new algorithms into a common framework and show their connections with the classical theory.

Finally, Chapter 5 deals with the application of feedforward neural networks to speech processing. A speech signal is the most fundamental communication medium, and it is also a typical example of dynamic (temporal and nonstationary) and nonlinear signals, which are usually difficult to handle in traditional system frameworks. To alleviate such difficulty, extensive research efforts have been expended on the application of feedforward networks to speech processing. Specifically, Chapter 5 starts by summarizing speech-related techniques and reviewing feedforward neural networks from the viewpoint of fundamental design issues such as the selection of network structures and the selection of training objective functions. We specially feature the recent design framework called the generalized probabilistic descent method in order to provide a comprehensive perspective about the issues involved in speech processing. We discuss the topic of speech recognition, to which feedforward neural networks have been most extensively applied. Other topics are summarized in an archived form. Through considerations of design fundamentals and application examples, the reader is enabled to understand the key points in the design of feedforward neural-network-based speech processing systems, the importance of special mechanism of shift tolerance and state transition.

The idea to write this volume came from Simon Haykin who selected the authors.

IRWIN W. SANDBERG
JAMES T. LO
CRAIG L. FANCOURT
JOSE C. PRINCIPE
SHIGERU KATAGIRI
SIMON HAYKIN

August 2000

NONLINEAR
DYNAMICAL SYSTEMS

1

FEEDFORWARD NEURAL NETWORKS: AN INTRODUCTION

Simon Haykin

A *neural network* is a massively parallel distributed processor that has a natural propensity for storing experiential knowledge and making it available for use. It resembles the brain in two respects (Haykin 1998):

1. Knowledge is acquired by the network through a learning process.
2. Interconnection strengths known as synaptic weights are used to store the knowledge.

Basically, learning is a process by which the free parameters (i.e., synaptic weights and bias levels) of a neural network are adapted through a continuing process of stimulation by the environment in which the network is embedded. The type of learning is determined by the manner in which the parameter changes take place. In a general sense, the learning process may be classified as follows:

- Learning with a teacher, also referred to as supervised learning
- Learning without a teacher, also referred to as unsupervised learning

1.1 SUPERVISED LEARNING

This form of learning assumes the availability of a labeled (i.e., ground-truthed) set of training data made up of N input—output examples:

$$T = \{(\mathbf{x}_i, d_i)\}_{i=1}^{N} \tag{1.1}$$

where $\mathbf{x}_i$ = input vector of ith example
  d_i = desired (target) response of ith example, assumed to be scalar for convenience of presentation
  N = sample size

Given the training sample T, the requirement is to compute the free parameters of the neural network so that the actual output y_i of the neural network due to $\mathbf{x}_i$ is close enough to d_i for all i in a statistical sense. For example, we may use the mean-square error

$$E(n) = \frac{1}{N} \sum_{i=1}^{N} (d_i - y_i)^2 \tag{1.2}$$

as the index of performance to be minimized.

1.1.1 Multilayer Perceptrons and Back-Propagation Learning

The back-propagation algorithm has emerged as the workhorse for the design of a special class of layered feedforward networks known as *multilayer perceptrons* (MLP). As shown in Fig. 1.1, a multilayer perceptron has an input layer of source nodes and an output layer of neurons (i.e., computation nodes); these two layers connect the network to the outside world. In addition to these two layers, the multilayer perceptron usually has one or more layers of hidden neurons, which are so called because these neurons are not directly accessible. The hidden neurons extract important features contained in the input data.

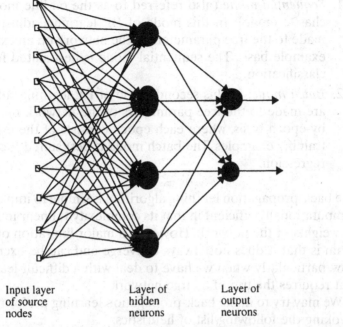

Input layer of source nodes

Layer of hidden neurons

Layer of output neurons

Figure 1.1 Fully connected feedforward with one hidden layer and one output layer.

The training of an MLP is usually accomplished by using a *back-propagation (BP) algorithm* that involves two phases (Werbos 1974; Rumelhart et al. 1986):

- *Forward Phase*. During this phase the free parameters of the network are fixed, and the input signal is propagated through the network of Fig. 1.1 layer by layer. The forward phase finishes with the computation of an error signal

$$e_i = d_i - y_i \qquad (1.3)$$

where d_i is the desired response and y_i is the actual output produced by the network in response to the input $\mathbf{x}_i$.

- *Backward Phase*. During this second phase, the error signal e_i is propagated through the network of Fig. 1.1 in the backward direction, hence the name of the algorithm. It is during this phase that adjustments are applied to the free parameters of the network so as to minimize the error e_i in a statistical sense.

Back-propagation learning may be implemented in one of two basic ways, as summarized here:

1. *Sequential mode* (also referred to as the on-line mode or sto-chastic mode): In this mode of BP learning, adjustments are made to the free parameters of the network on an example-by-example basis. The sequential mode is best suited for pattern classification.

2. *Batch mode*: In this second mode of BP learning, adjustments are made to the free parameters of the network on an epoch-by-epoch basis, where each epoch consists of the entire set of training examples. The batch mode is best suited for nonlinear regression.

The back-propagation learning algorithm is simple to implement and computationally efficient in that its complexity is linear in the synaptic weights of the network. However, a major limitation of the algorithm is that it does not always converge and can be excruciatingly slow, particularly when we have to deal with a difficult learning task that requires the use of a large network.

We may try to make back-propagation learning perform better by invoking the following list of heuristics:

- Use neurons with antisymmetric activation functions (e.g., hyperbolic tangent function) in preference to nonsymmetric activation functions (e.g., logistic function). Figure 1.2 shows examples of these two forms of activation functions.

- Shuffle the training examples after the presentation of each epoch; an epoch involves the presentation of the entire set of training examples to the network.

- Follow an easy-to-learn example with a difficult one.

- Preprocess the input data so as to remove the mean and decorrelate the data.

- Arrange for the neurons in the different layers to learn at essentially the same rate. This may be attained by assigning a learning rate parameter to neurons in the last layers that is smaller than those at the front end.

- Incorporate prior information into the network design whenever it is available.

One other heuristic that deserves to be mentioned relates to the size of the training set, N, for a pattern classification task. Given a multilayer perceptron with a total number of synaptic weights including bias levels, denoted by W, a rule of thumb for selecting N is

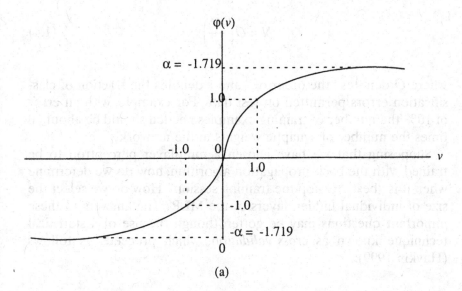

(a)

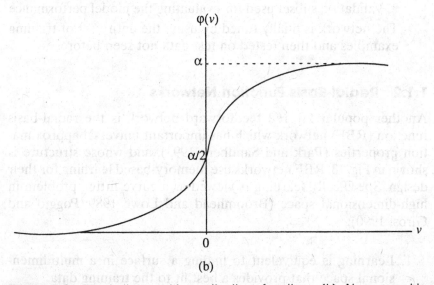

(b)

Figure 1.2 (a) Antisymmetric activation function. (b) Nonsymmetric activation function.

$$N = O\left(\frac{W}{\varepsilon}\right) \tag{1.4}$$

where O denotes "the order of," and ε denotes the fraction of classification errors permitted on test data. For example, with an error of 10% the number of training examples needed should be about 10 times the number of synaptic weights in the network.

Supposing that we have chosen a multilayer perceptron to be trained with the back-propagation algorithm, how do we determine when it is "best" to stop the training session? How do we select the size of individual hidden layers of the MLP? The answers to these important questions may be gotten though the use of a statistical technique known as *cross-validation*, which proceeds as follows (Haykin 1999):

- The set of training examples is split into two parts:
  - Estimation subset used for training of the model
  - Validation subset used for evaluating the model performance
- The network is finally tuned by using the entire set of training examples and then tested on test data not seen before.

1.1.2 Radial-Basis Function Networks

Another popular layered feedforward network is the radial-basis function (RBF) network which has important universal approximation properties (Park and Sandberg 1993), and whose structure is shown in Fig. 13. RBF networks use memory-based learning for their design. Specifically, learning is viewed as a curve-fitting problem in high-dimensional space (Broomhead and Lowe 1989; Poggio and Girosi 1990):

1. Learning is equivalent to finding a surface in a multidimensional space that provides a best fit to the training data.
2. Generalization (i.e., response of the network to input data not seen before) is equivalent to the use of this multidimensional surface to interpolate the test data.

RBF networks differ from multilayer perceptrons in some fundamental respects:

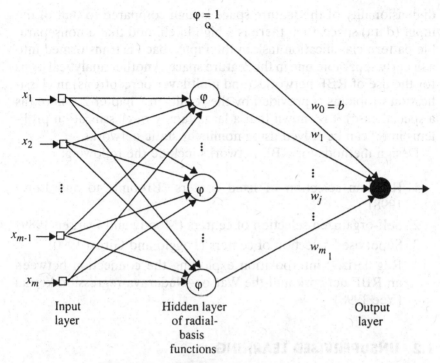

Figure 1.3 Radial-basis function network.

- RBF networks are local approximators, whereas multilayer perceptrons are global approximators.
- RBF networks have a single hidden layer, whereas multilayer perceptrons can have any number of hidden layers.
- The output layer of a RBF network is always linear, whereas in a multilayer perceptron it can be linear or nonlinear.
- The activation function of the hidden layer in an RBF network computes the Euclidean distance between the input signal vector and parameter vector of the network, whereas the activation function of a multilayer perceptron computes the inner product between the input signal vector and the pertinent synaptic weight vector.

The use of a linear output layer in an RBF network may be justified in light of *Cover's theorem* on the separability of patterns. According to this theorem, provided that the transformation from the input space to the feature (hidden) space is nonlinear and the dimensionality of the feature space is high compared to that of the input (data) space, then there is a high likelihood that a nonseparable pattern classification task in the input space is transformed into a linearly separable one in the feature space. Another analytical basis for the use of RBF networks (and multilayer perceptrons) in classification problems is provided by the results in Chapter 2, where (as a special case) it is shown that a large family of classification problems in $\mathbb{R}^n$ can be solved using nonlinear static networks.

Design methods for RBF networks include the following:

1. Random selection of fixed centers (Broomhead and Lowe 1998)
2. Self-organized selection of centers (Moody and Darken 1989)
3. Supervised selection of centers (Poggio and Girosi 1990)
4. Regularized interpolation exploiting the connection between an RBF network and the Watson–Nadaraya regression kernel (Yee 1998).

1.2 UNSUPERVISED LEARNING

Turning next to unsupervised learning, adjustment of synaptic weights may be carried through the use of neurobiological principles such as Hebbian learning and competitive learning. In this section we will describe specific applications of these two approaches.

1.2.1 Principal Components Analysis

According to *Hebb's postulate of learning*, the change in synaptic weight Δw_{ji} of a neural network is defined by

$$\Delta w_{ji} = \eta x_i y_j \tag{1.5}$$

where η = learning-rate parameter
x_i = input (presynaptic) signal
y_i = output (postsynaptic) signal

Principal component analysis (PCA) networks use a modified form of this self-organized learning rule. To begin with, consider a linear neuron designed to operate as a maximum eigenfilter; such a neuron is referred to as *Oja's neuron* (Oja 1982). It is characterized as follow:

$$\Delta w_{ji} = \eta y_j (x_i - y_j w_{ji}) \tag{1.6}$$

where the term $-\eta y_j^2 w_{ji}$ is added to stabilize the learning process. As the number of iterations approaches infinity, we find the following:

1. The synaptic weight vector of neuron j approaches the eigenvector associated with the largest eigenvalue λ_{max} of the correlation matrix of the input vector (assumed to be of zero mean).
2. The variance of the output of neuron j approaches the largest eigenvalue λ_{max}.

The generalized Hebbian algorithm (GHA), due to Sanger (1989), is a straightforward generalization of Oja's neuron for the extraction of any desired number of principal components.

1.2.2 Self-Organizing Maps

In a self-organizing map (SOM), due to Kohonen (1997), the neurons are placed at the nodes of a lattice, and they become selectively tuned to various input patterns (vectors) in the course of a competitive learning process. The process is characterized by the formation of a topographic map in which the spatial locations (i.e., coordinates) of the neurons in the lattice correspond to intrinsic features of the input patterns. Figure 1.4 illustrates the basic idea of a self-organizing map, assuming the use of a two-dimensional lattice of neurons as the network structure.

In reality, the SOM belongs to the class of vector-coding algorithms (Luttrell, 1989). That is, a fixed number of codewords are placed into a higher-dimensional input space, thereby facilitating data compression.

An integral feature of the SOM algorithm is the neighborhood function centered around a neuron that wins the competitive process. The neighborhood function starts by enclosing the entire lattice initially and is then allowed to shrink gradually until it encompasses the winning neuron.

The algorithm exhibits two distinct phases in its operation:

1. *Ordering phase*, during which the topological ordering of the weight vectors takes place
2. *Convergence phase*, during which the computational map is fine tuned

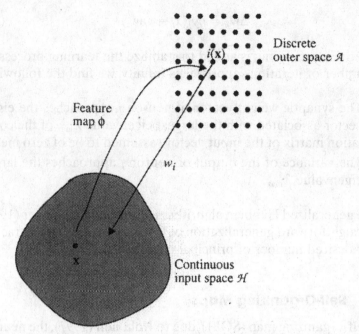

Figure 1.4 Illustration of relationship between feature map ϕ and weight vector $\mathbf{w}_i$ of winning neuron i.

The SOM algorithm exhibits the following properties:

1. Approximation of the continuous input space by the weight vectors of the discrete lattice.
2. Topological ordering exemplified by the fact that the spatial location of a neuron in the lattice corresponds to a particular feature of the input pattern.
3. The feature map computed by the algorithm reflects variations in the statistics of the input distribution.
4. SOM may be viewed as a nonlinear form of principal components analysis.

Refinements to the SOM algorithm are discussed in Van Hulle (2000).

1.3 TEMPORAL PROCESSING USING FEEDFORWARD NETWORKS

The material just described is concerned basically with the approximation of systems without dynamics (i.e., with static systems). At about the same time as the appearance of the early universal-approximation theorems for static neural networks there began (Sandberg 1991a) a corresponding study (see Chapter 2) of the neural network approximation of approximately-finite-memory maps and myopic maps. It was found that large classes of these maps can be uniformly approximated arbitrarily well by the maps of certain simple nonlinear structures using, for example, sigmoidal nonlinearities or radial basis functions. The approximating networks are two-stage structures comprising a linear preprocessing stage followed by a memoryless nonlinear network. Much is now known about the properties of these networks, and examples of these properties are given in the following.

From another perspective, *time* is an essential dimension of learning. We may incorporate time into the design of a neural network implicitly or explicitly. A straightforward method of implicit representation of time[1] is to add a *short-term memory structure* in the input layer of a static neural network (e.g., multilayer perceptron). The resulting configuration is sometimes called a *focused time-lagged feedforward network (TLFN)*.

The short-term memory structure may be implemented in one of two forms, as described here:

1. *Tapped-Delay-Line (TDL) Memory*. This is the most commonly used form of short-term memory. It consists of p unit delays with $(p + 1)$ terminals, as shown in Fig. 1.5, which may be viewed as a single input–multiple output network. Figure 1.6 shows a focused TLFN network using the combination of a TDL memory and multilayer perceptron. In Figs. 1.5 and 1.6, the unit-delay is denoted by z^{-1}.

   The *memory depth* of a TDL memory is fixed at p, and its *memory resolution* is fixed at unity, giving a *depth resolution constant* of p.

2. *Gamma Memory*. We may exercise control over the memory depth by building a feedback loop around each unit delay, as illustrated in Fig. 1.7 (de Vries and Principe 1992). In effect, the unit delay z^{-1} of the standard TDL memory is replaced by the transfer function

---

[1] Another practical way of accounting for time in a neural network is to employ feedback at the local or global level. Neural networks so configured are referred to as recurrent networks. For detailed treatment of recurrent networks, see Haykin (1999).

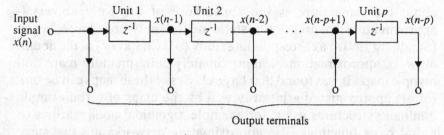

Figure 1.5 Ordinary tapped-delay line memory of order p.

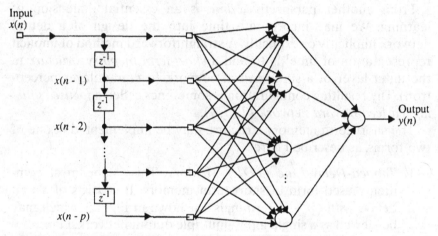

Figure 1.6 Focused time-lagged feedforward network (TLFN); the bias levels have been omitted for convenience of presentation.

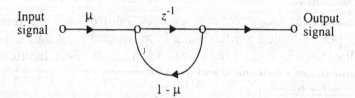

Figure 1.7 Signal-flow graph for one section of gamma memory.

$$G(z) = \frac{\mu z^{-1}}{1 - (1 - \mu)z^{-1}}$$

$$= \frac{\mu}{z - (1 - \mu)}$$

where μ is an adjustable parameter. For stability, the only pole of $G(z)$ at $z = (1 - \mu)$ must lie inside the unit circle in the z plane. This, in turn, requires that we restrict the choice of μ to the following range of values:

$$0 < \mu < 2$$

The overall impulse response of the gamma memory, consisting of p sections, is the inverse z transform of the overall transfer function

$$G_p(z) = \left(\frac{\mu}{z - (1 - \mu)} \right)^p$$

Denoting the impulse response by $g_p(n)$, we have

$$g_p(z) = \binom{n-1}{p-1} \mu^p (1 - \mu)^{n-p} \qquad n \geq p$$

where (:) is a binomial coefficient. The overall impulse response $g_p(n)$ for varying p represents a discrete version of the integrand of the *gamma function* (deVries and Principe 1992); hence the name "gamma memory."

The depth of the gamma memory is p/μ and its resolution is μ, for a depth resolution product of p. Accordingly, by choosing μ to be less than unity, the gamma memory provides improvement in depth over the TDL memory, but at the expense of memory resolution.

With regard to the utility of gamma networks—which are particular cases of the family of two-stage structures comprising a linear preprocessing stage followed by a memoryless nonlinear network—experimental results have been reported which indicate that the structure is useful. In fact, it is known (Sandberg and Xu 1997) that for a large class of discrete-time dynamic system maps H, and for any choice of μ in the interval $(0, 1)$, there is a focused gamma network that approximates H uniformly arbitrarily well. It is known that tapped-delay-line networks (i.e., networks with $\mu = 1$) also have the universal approximation property (Sandberg 1991b).

Focused TLFNs using ordinary tapped-delay memory or gamma memory are limited to stationary environments. To deal with nonstationary dynamical processes, we may use distributed TLFNs where the effect of time is distributed at the synaptic level throughout the network. One way in which this may be accomplished is to use finite-duration impulse response (FIR) filters to implement the synaptic connections of an MLP; Fig. 1.8 shows an FIR model of a synapse. The training of a distributed TLFN is naturally a more difficult proposition than the training of a focused TLFN. Whereas we may use the ordinary back-propagation algorithm to train a focused TLFN, we have to extend the back-propagation algorithm to cope with the replacement of a synaptic weight in the ordinary MLP by a synaptic weight vector. This extension is referred to as the temporal back-propagation algorithm due to Wan (1994).

1.4 CONCLUDING REMARKS

In this chapter, we briefly reviewed the feedforward type of neural networks, which are exemplified by multilayer perceptrons (MLPs), radial-basis function (RBF) networks, principal component analysis (PCA) networks, and self-organizing maps (SOMs). The training of MLPs and RBF networks proceeds in a supervised manner, whereas the training of PCA networks and SOMs proceeds in an unsupervised manner.

Feedforward networks by themselves are nonlinear static networks. They can be made to operate as nonlinear dynamical systems

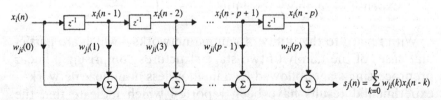

Figure 1.8 Finite-duration impulse response (FIR) filter.

by incorporating short-term memory into their input layer. Two important examples of short-term memory are the standard tapped-delay-line and the gamma memory that provides control over attainable memory depth. The attractive feature of nonlinear dynamical systems built in this way is that they are inherently stable.

BIBLIOGRAPHY

Anderson, J. A., 1995, *Introduction to Neural Networks* (Cambridge, MA: MIT Press).

Barlow, H. B., 1989, "Unsupervised learning," *Neural Computation*, vol. 1, pp. 295–311.

Becker, S., and G. E. Hinton, 1982, "A self-organizing neural network that discovers surfaces in random-dot stereograms," *Nature (London)*, vol. 355, pp. 161–163.

Broomhead, D. S., and D. Lowe, 1988, "Multivariable functional interpolation and adaptive networks," *Complex Systems*, vol. 2, pp. 321–355.

Comon, P., 1994, "Independent component analysis: A new concept?" *Signal Processing*, vol. 36, pp. 287–314.

deVries, B., and J. C. Principe, 1992, "The gamma model—A new neural model for temporal processing," *Neural Networks*, vol. 4, pp. 565–576.

Haykin, S., 1999, *Neural Networks: A Comprehensive Foundation*, 2nd ed. (Englewood Cliffs, NJ: Prentice-Hall).

Moody and Darken, 1989, "Fast learning in networks of locally-tuned processing unites," *Neural Computation*, vol. 1, pp. 281–294.

Oja, E., 1982, "A simplified neuron model as a principal component analyzer," *J. Math. Biol.*, vol. 15, pp. 267–273.

Park, J., and Sandberg, I. W., 1993, "Approximation and radial-basis function networks," *Neural computation*, vol. 5, pp. 305–316.

Poggio, T., and F. Girosi, 1990, "Networks for approximation and learning," *Proc. IEEE*, vol. 78, pp. 1481–1497.

Rumelhart, D. E., G. E. Hinton, and R. J. Williams, 1986, "Learning internal representations by error propagation," in D. E. Rumelhart and J. L. McCleland, eds. (Cambridge, MA: MIT Press), vol. 1, Chapter 8.

Sandberg, I. W., 1991a, "Structure theorems for nonlinear systems," *Multidimensional Sys. Sig. Process.* vol. 2, pp. 267–286. (Errata in 1992, vol. 3, p. 101.)

Sandberg, I. W., 1991b, "Approximation theorems for discrete-time systems," *IEEE Trans. Circuits Sys.* vol. 38, no. 5, pp. 564–566, May 1991.

Sandberg, I. W., and Xu, L., 1997, "Uniform approximation and gamma networks," *Neural Networks*, vol. 10, pp. 781–784.

Van Hulle, M. M., 2000, *Faithful Representations and Topographic Maps: From Distortion-to-Information-Based Self Organization* (New York: Wiley).

Wan, E. A., 1994, "Time series prediction by using a connectionist network with internal delay lines," in A. S. Weigend and N. A. Gershenfield, eds., *Time Series Prediction: Forecasting the Future and Understanding the Past* (Reading, MA: Addison-Wesley), pp. 195–217.

Werbos, P. J., 1974, "Beyond regression: New tools for prediction and analysis in the behavioral sciences," Ph.D. Thesis, Harvard University, Cambridge, MA.

Werbos, P. J., 1990, "Backpropagation through time: What it does and how to do it," *Proc. IEEE*, vol. 78, pp. 1550–1560.

Yee, P. V., 1998, "*Regularized radial basis function networks: Theory and applications to probability estimation, classification, and time series prediction*," Ph.D. Thesis, McMaster University, Hamilton, Ontario.

2

UNIFORM APPROXIMATION AND NONLINEAR NETWORK STRUCTURES

Irwin W. Sandberg

2.1 INTRODUCTION

In this chapter we consider classification problems and the related problem of approximating dynamic nonlinear input output maps. Attention is focused on the properties of nonlinear structures that have the form of a dynamic preprocessing stage followed by a memoryless nonlinear section.

The problem of classifying signals is of interest in several application areas. Typically, we are given a finite number m of pairwise disjoint sets $C_1, \ldots, C_m$ of signals, and we would like to synthesize a system that maps the elements of each C_j into a real number a_j, such that the numbers $a_1, \ldots, a_m$ are distinct. One of the main purposes of this chapter is to show that this classification can be performed by certain simple structures, involving linear functionals and memoryless nonlinear elements, assuming only that the C_j are compact subsets of a real normed linear space. The results on which this conclusion is based, which are given in Section 2.2, have applications other than to classification problems. For example, one result provides a simplification of a proof of a well-known proposition concerning approximations in $\mathbb{R}^n$ using sigmoidal functions. Other material concerning the general problem of classification can be found in, for example, Haykin (1999, Chapters 4–6).

In Section 2.3 we consider the problem of approximating discrete-space multidimensional shift-invariant input–output maps with vector-valued inputs drawn from a certain large set. Our main result there is a theorem showing that if the maps to be approximated satisfy a certain "myopicity" condition, which is very often met, then they can be uniformly approximated arbitrarily well[1] using a structure consisting of a linear preprocessing stage followed by a memoryless nonlinear network. [Roughly speaking, an input–output map G is "myopic" if the value of $(Gx)(\alpha)$ is always relatively independent of the values of x at points remote from α.] Noncausal as well as causal maps are considered in Section 2.3. Approximations for noncausal maps for which inputs and outputs are functions of more than one variable are of current interest in connection with, for example, image processing.

In Section 2.4 we consider a different aspect of the problem of approximating nonlinear input–output maps: While in Section 2.3 the preprocessing stage is assumed at the outset to satisfy certain key conditions, in Section 2.4 the focus is on obtaining an understanding of the basic properties that the preprocessing stage must possess. In other words, in Section 2.4 we are interested in characterizing those preprocessing stages that permit uniform approximation. There are also other differences in Section 2.4 where the definition of a myopic map is different than in Section 2.3 (but the main idea is similar), and an important role is played by the related concept of maps that possess "approximately finite memory" in a setting in which inputs and outputs are defined on the nonnegative integers. We first consider causal time-invariant nonlinear maps that take a set of bounded functions into a set of real-valued functions, and we give criteria under which these maps can be uniformly approximated arbitrarily

---

[1] The term *uniformly approximated* refers to approximation with respect to some sup (i.e., least upper bound) norm—as opposed to, for example, approximation in the mean. For instance, the statement that any real-valued continuous function f defined on $[0, 1]$ can be uniformly approximated arbitrarily well by a real polynomial means that for any such f and for each $\varepsilon > 0$ there is a real polynomial p such that $\max_{x \in [0,1]} |f(x) - p(x)| < \varepsilon$. Often "uniformly approximated" is used to mean "uniformly approximated arbitrarily well."

well using a structure consisting of a not-necessarily linear dynamic part followed by a nonlinear memoryless section that may contain sigmoids or radial basis functions, etc. In our results certain separation conditions, of the kind associated with the Stone–Weierstrass theorem, play a prominent role. Here they emerge as criteria for approximation and not just sufficient conditions under which an approximation exists. As an application of the results, we show that system maps of the type addressed can be uniformly approximated arbitrarily well by doubly finite Volterra series approximants *if and only if* these maps have approximately finite memory and satisfy certain continuity conditions. Corresponding results are then given for (not necessary causal) multivariable input–output maps. As mentioned above, such multivariable maps are of interest in connection with image processing. Further introductory material is given in Sections 2.2 to 2.4.

It will become clear that we make use of concepts drawn from the areas of real analysis and functional analysis. These are useful powerful ideas. Since not all readers interested in the topic of this chapter are familiar with these ideas, we mention that there are many very good pertinent references. See, for example, Kolmogorov and Fomin (1970), Natanson (1960), and Mukherjea and Pothoven (1978).

2.2 GENERAL STRUCTURES FOR CLASSIFICATION

The problem of classifying signals is of interest in several application areas (e.g., signal detection, computer-assisted medical diagnosis, automatic target identification, etc.). Typically we are given a finite number m of pairwise disjoint sets $C_1, \ldots, C_m$ of signals, and we would like to synthesize a system that maps the elements of each C_j into a real number a_j so that the numbers $a_1, \ldots, a_m$ are distinct. One purpose of the next section is to show that the structure shown in Fig. 2.1 can perform this classification (for any prescribed distinct $a_1, \ldots, a_m$) assuming only that the C_j are compact subsets of a real normed linear space X. In the figure, x is the signal to be classified, $y_1, \ldots, y_k$ are functionals that can be taken to be linear,[2] h denotes a continuous memoryless nonlinear mapping that, for example, can be implemented by a neural network having one hidden nonlinear layer, the block labeled Q is a quantizer that for each j maps numbers

---

[2] For example, if X is $L_2(a, b)$ the functionals can be taken to be inner-product operators. Other examples are given in Section 2.2.2.

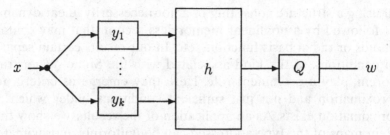

Figure 2.1 Classification network.

in the interval $(a_j - 0.25\rho, a_j + 0.25\rho)$ into a_j, where $\rho = \min_{i \neq j}|a_i - a_j|$, and w is the output of the classifier.

A proof is given in the next section.[3] [Take ε in part (ii) of Corollary 1 there to be less than 0.25ρ.] The proof is an existence proof; the problem of determining the elements of the structure of Fig. 2.1 is not considered in this chapter. Since h is continuous and the quantizer map can be assumed to be continuous, notice that the result outlined above establishes the proposition that classifiers of the type considered can be realized as a cascade of a system that realizes linear functionals followed by a system that implements a continuous memoryless map.

It is also shown in Section 2.2.1 that h can be synthesized as a parallel combination of simple structures followed by a summing device, as shown in Fig. 2.2. In the figure each block labeled c_j denotes multiplication by a real number c_j, and every u_j block denotes the taking of a real-valued function u_j defined on the reals. The u_j are drawn from a set U of continuous functions whose finite sums can approximate the exponential function $\exp(\cdot)$ on bounded intervals. In particular, u_j can be taken to be $\exp(\cdot)$ for all j. A classification example is given in Section 2.2.3.

The results in Section 2.2.1 have applications other than those described above. For example, one result provides a simplification of a proof of a well-known proposition concerning approximations in $\mathbb{R}^n$ using sigmoidal functions (see Section 2.2.4).

---

[3] Theorem 1 in the next section expands on a brief remark in Sandberg (1991, p. 274) that any continuous real functional on a compact subset of a real normed linear space can be uniformly approximated arbitrarily well using only a feedforward neural network with a linear functional input layer and one memoryless hidden nonlinear (e.g., sigmoidal) layer. The theorem in Section 2.2.1 is considerably more usful because of additional flexibility concerning h and the functionals $y_1, \ldots, y_k$. Some of the results in this chapter are described briefly in Sandberg (1993a).

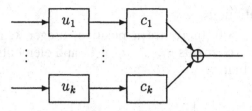

Figure 2.2 Structure for h.

2.2.1 Theorem 1: A Characterization of Continuity

Let C be a nonempty compact subset of a real normed linear space X, and let X^* be the set of bounded linear functionals on X (i.e., the set of bounded linear maps from X to the reals $\mathbb{R}$). Let Y be any set of continuous maps from X to $\mathbb{R}$ that is dense in X^* on C, in the sense that for each $\phi \in X^*$ and any $\varepsilon > 0$ there is a $y \in Y$ such that $|\phi(x) - y(x)| < \varepsilon, x \in C$. For example, a theorem in Section 2.2.2 shows that functionals y of the form $y(x) = \sum_{i=1}^{p} c_i x(\alpha_i)$ where the c_i belong to $\mathbb{R}$ and $\alpha_i \in [0, 1]^n$ for each i, and also functionals y of the form $y(x) = \int_{[0,1]^n} q(\alpha) x(\alpha) \, d\alpha$ where $q: [0, 1]^n \to \mathbb{R}$ is continuous, are dense in X^* on C for X the set of $\mathbb{R}$-valued continuous functions on $[0, 1]^n$ with the usual max norm.

For each $k = 1, 2, \ldots$ let D_k denote any family of continuous maps $h: \mathbb{R}^k \to \mathbb{R}$ such that given a compact $E \subset \mathbb{R}^k$ and any continuous $g: E \to \mathbb{R}$ as well as any $\sigma > 0$ there is an $h \in D_k$ such that $|g(x) - h(x)| < \sigma$ for $x \in E$.[4] Finally, let U be any set of continuous maps $u: \mathbb{R} \to \mathbb{R}$ such that given $\sigma > 0$ and any bounded interval $(\beta_1, \beta_2) \subset \mathbb{R}$ there exists a finite number of elements $u_1, \ldots, u_\ell$ of U for which $|\exp(\beta) - \sum_j u_j(\beta)| < \sigma$ for $\beta \in (\beta_1, \beta_2)$.[5]

Theorem 1. Assuming that f maps C into $\mathbb{R}$, the following three conditions are equivalent.

---

[4] For example, the elements of each D_k can be represented by one-hidden-layer neural networks with sigmoidal nonlinearities, or by radial-basis-function networks, or by polynomial networks, etc.

[5] Of course, we can take U to be the set whose only element is $\exp(\cdot)$, or the set $\{u: u(\beta) = (\beta)^n/n!, n \in \{0, 1, \ldots\}\}$. Another acceptable choice is $\{u: u(\beta) = cs(w\beta + \rho), c, w, \rho \in \mathbb{R}\}$, where s is a continuous function with $\lim_{\beta \to \infty} s(\beta) = 1$ and $\lim_{\beta \to -\infty} s(\beta) = 0$ (Cybenko 1989).

(i) f is continuous.
(ii) Given $\varepsilon > 0$, there are a positive integer k, real numbers $c_1, \ldots, c_k$, elements $u_1, \ldots, u_k$ of U, and elements $y_1, \ldots, y_k$ of Y such that

$$\left| f(x) - \sum_j c_j u_j[y_j(x)] \right| < \varepsilon$$

for $x \in C$.
(iii) Given $\varepsilon > 0$ there are a positive integer k, elements $y_1, \ldots, y_k$ of Y, and an $h \in D_k$ such that

$$|f(x) - h[y_1(x), \ldots, y_k(x)]| < \varepsilon$$

for $x \in C$.

Proof. Suppose first that (i) is met and notice that the set V of all functions $v: C \to \mathbb{R}$ of the form

$$v(x) = \sum_j a_j \exp[\phi_j(x)],$$

in which the sum is finite and the a_j and the ϕ_j belong to $\mathbb{R}$ and X^*, respectively, is an algebra under the natural definition of addition and multiplication. By a consequence (Bachman and Narici 1966, p. 198) of the Hahn–Banach theorem,[6] given distinct x_a and x_b in C, there is a ϕ in X^* such that $\exp[\phi(x_a)] \neq \exp[\phi(x_b)]$, showing that V separates the points of C. It is clear that $v(x) \neq 0$ for some $v \in V$ for each x. Thus, by a version of the Stone–Weierstrass theorem (Rudin 1976, p. 162),[7] given $\varepsilon > 0$ there are a positive integer p, real numbers $d_1, \ldots, d_p$, and elements $z_1, \ldots, z_p$ of X^* such that

---

[6] This consequence of the Hahn–Banach theorem tells us that if x_0 is a nonzero element of X, then X^* contains an F such that $\|F\| = 1$ and $F(x_0) = \|x_0\|$.
[7] According to this theorem, if $\mathcal{A}$ is an algebra of continuous real-valued functions on a compact set K, if $\mathcal{A}$ separates the points of K and does not vanish throughout K, then the uniform closure of $\mathcal{A}$ consists of all continuous real-valued functions on K.

$$\left| f(x) - \sum_j d_j \exp[z_j(x)] \right| < \frac{\varepsilon}{3} \qquad (2.1)$$

for $x \in C$.[8]

We may assume that $\Sigma_j|d_j| \neq 0$. Choose $\gamma > 0$ such that $\gamma\Sigma_j|d_j| < \varepsilon/3$. Let $[a', b']$ be an interval in $\mathbb{R}$ that contains all of the sets $z_j(C)$, and let real a and b be such that $a < a'$, $b > b'$. Select $\eta > 0$ such that $|\exp(\beta_1) - \exp(\beta_2)| < \gamma$ for $\beta_1, \beta_1 \in [a, b]$ with $|\beta_1 - \beta_2| < \eta$. With $\rho = \min(\eta, a' - a, b - b')$, choose $y_i \in Y$ such that $|z_j(x) - y_i(x)| < \rho$, $x \in C$ for all j. This gives $|\exp[z_j(x)] - \exp[y_i(x)]| < \gamma$, $x \in C$ for each j [because we have $y_j(C) \in [a, b]$ and $|z_j(x) - y_i(x)| < \eta$ for each j and x], and thus

$$\left| f(x) - \sum_j d_j \exp[y_j(x)] \right|$$

$$\leq \left| f(x) - \sum_j d_j \exp[z_j(x)] \right| + \left| \sum_j d_j \exp[z_j(x)] - \sum_j d_j \exp[y_j(x)] \right|$$

$$\leq \frac{\varepsilon}{3} + \sum_j |d_j| \cdot |\exp[z_j(x)] - \exp[y_j(x)]| \leq \frac{2\varepsilon}{3}, \qquad x \in C$$

Now pick $u_1, \ldots, u_\ell \in U$ so that

$$\left| \exp(\beta) - \sum_i u_i(\beta) \right| \leq \gamma_1, \qquad \beta \in [a, b]$$

where $\gamma_1\Sigma_j|d_j| < \varepsilon/3$. Then, since $y_i(C) \subset [a, b]$ for each j,

$$\left| f(x) - \sum_j \sum_i d_j u_i[y_j(x)] \right| \leq \left| f(x) - \sum_j d_j \exp[y_j(x)] \right| + \left| \sum_j d_j \exp[y_j(x)] \right.$$

$$\left. - \sum_j \sum_i d_j u_i[y_j(x)] \right| \leq \frac{2\varepsilon}{3} + \sum_j \left| d_j \exp[y_j(x)] - d_j \sum_i u_i[y_i(x)] \right|$$

$$\leq \frac{2\varepsilon}{3} + \sum_j |d_j| \cdot \left| \exp[y_j(x)] - \sum_i u_i[y_i(x)] \right| \leq \frac{2\varepsilon}{3} + \gamma_1 \sum_j |d_j| < \varepsilon.$$

---

[8] Here we view C as a metric space with the metric derived in the usual way from the norm in X.

Since $\Sigma_j\Sigma_i d_j u_i[y_i(x)]$ can be written in the form $\Sigma_j c_j u_j[y_j(x)]$, with the c_j, u_j, and y_j in $\mathbb{R}$, U, and Y, respectively, (ii) holds.

To show that (ii) $\Rightarrow$ (iii), let $\varepsilon > 0$ and suppose that there are k, $c_1,\ldots,c_k$, and $u_1,\ldots,u_k$ as indicated such that

$$\left| f(x) - \sum_j c_j u_j[y_j(x)] \right| < \frac{\varepsilon}{2}, \qquad x \in C$$

Let $h \in D_k$ satisfy $|h(\lambda) - \Sigma_j c_j u_j(\lambda_j)| < \varepsilon/2$ for $\lambda \in [a, b]^k$. Then

$$|f(x) - h[y_1(x),\ldots,y_k(x)]| \leq \left| f(x) - \sum_j c_j u_j[y_j(x)] \right|$$
$$+ \left| \sum_j c_j u_j[y_j(x)] - h[y_1(x),\ldots,y_k(x)] \right| < \frac{\varepsilon}{2} + \frac{\varepsilon}{2}$$

for $x \in C$, as required. (iii) $\Rightarrow$ (i) is a consequence of the fact that f is the uniform limit of a sequence of continuous functions under (iii). This completes the proof.

Part (iii) of the theorem shows that any continuous real functional on a compact subset of a real normed linear space can be realized approximately by a cascade of a system that realizes $[y_1,\ldots,y_k]$ followed by a nonlinear system that realizes h.[9] It generalizes an early result due to Fréchet [see Volterra (1959, p. 20)] concerning the approximation of functionals on compact subsets C of continuous $\mathbb{R}$-valued functions defined on a finite interval $[a, b]$. [Fréchet's result is obtained from the theorem (with C as indicated here) by taking the D_k to be families of polynomial maps and taking the elements y of Y to be of the form $y(x) = \int_a^b q(t)x(t)\,dt$, where $q: [a, b] \to \mathbb{R}$ is continuous.] Part (ii) of the theorem shows that h can always be taken to be a map of a certain specific form.

Corollary 1. Let $C = C_1 \cup C_2 \cup \ldots \cup C_m$ (m any positive integer), where the C_j are pairwise disjoint compact subsets of X, let $\{a_1,\ldots,a_m\}$ be a set of real numbers, and let $f: C \to \mathbb{R}$ satisfy $f(x) = a_j$ for $x \in C_j$, $j = 1,\ldots,m$. Then given $\varepsilon > 0$, the following hold.

---

[9] For related different results, see Sandberg (1991, 1992a).

(i) There are a positive integer k, real numbers $c_1, \ldots, c_k$, elements $y_1, \ldots, y_k$ of Y, and elements $u_1, \ldots, u_k$ of U such that

$$a_j - \varepsilon < \sum_j c_j u_j [y_j(x)] < a_j + \varepsilon$$

for $x \in C_j$ and $j = 1, \ldots, m$.

(ii) There are a positive integer k, elements $y_1, \ldots, y_k$ of Y, and an $h \in D_k$ such that

$$a_j - \varepsilon < h[y_1(x), \ldots, y_k(x)] < a_j + \varepsilon$$

for $x \in C_j$ and $j = 1, \ldots, m$.

This follows from the theorem and the observation that f is continuous on C. (f is continuous because it is constant on each C_j and the distance between any pair of distinct nonempty C_j is positive, because the C_j are closed and pairwise disjoint.)[10]

2.2.2 Examples of Y

The following theorem provides two examples of the set Y.

Theorem 2. Let X denote the normed linear space of $\mathbb{R}$-valued continuous functions on $I := [0, 1]^n$, with the usual max norm. Let $g \in X^*$, and let $\varepsilon > 0$. Then there are points $\alpha_1, \ldots, \alpha_p \in I$, points $c_1, \ldots, c_p \in \mathbb{R}$, and a $q \in X$ such that

$$\sup_{x \in C} \left| g(x) - \sum_{j=1}^p c_j x(\alpha_j) \right| < \varepsilon \tag{2.2}$$

and

$$\sup_{x \in C} \left| g(x) - \int_I q(\alpha) x(\alpha) \, d\alpha \right| < \varepsilon \tag{2.3}$$

Proof. We will use the following lemma.

---

[10] We take this opportunity to correct an error in the corollary in Sandberg (1993a): $f(x) - h[y_1(x), \ldots, y_k(x)]$ should be replaced with $h[y_1(x), \ldots, y_k(x)]$.

Lemma 1. Let A and B be normed linear spaces, and let K be a compact subset of A. Let $0 < k_1 \leq k_2 < \infty$, and suppose that $G: K \to B$ belongs to $\text{Lip}(k_1)$ (i.e., has the property that $\|Gu - Gv\| \leq k_1\|u - v\|$ for $u, v \in K$).[11] Let $\mathcal{F}$ be any family of maps $H: K \to B$ such that the elements of $\mathcal{F}$ belong to $\text{Lip}(k_2)$ and such that given any $\delta > 0$ and any finite set $\{x_1, \ldots, x_r\}$ of points of K there is an $H \in \mathcal{F}$ for which

$$\|Gx_i - Hx_i\| < \delta \ (1 \leq i \leq r)$$

Then given $\varepsilon > 0$ there is an $H \in \mathcal{F}$ such that $\|Gx - Hx\| < \varepsilon$ for all $x \in K$.

Proof of the Lemma.

Let $\varepsilon > 0$ be given and (by the compactness of K) let $b_1(x_1), \ldots, b_r(x_r)$ be an open cover of K, with the $b(x_i)$ open balls of radius $\varepsilon/(3k_2)$ centered at $x_i \in K$. Let $H \in \mathcal{F}$ satisfy $\|Gx_i - Hx_i\| < \varepsilon/3$ for all i.

Choose $x \in K$. There is $i \in \{1, \ldots, r\}$ such that $\|x - x_i\| < \varepsilon/(3k_2)$. For this i,

$$\|Gx - Gx_i\| \leq k_1\|x - x_i\| < k_1\varepsilon/(3k_2) \leq \varepsilon/3$$

Also,

$$\|Gx - Hx\| \leq \|Gx - Gx_i\| + \|Gx_i - Hx_i\| + \|Hx_i - Hx\|$$
$$< \varepsilon/3 + \varepsilon/3 + k_2\varepsilon/(3k_2) = \varepsilon$$

which proves the lemma.[12]

Returning to the proof of the theorem, the functional g has the representation

$$g(x) = \int_I x \, d\mu$$

---

[11] We use the same symbol $\|\cdot\|$ for the norm in the two spaces. The elements of $\text{Lip}(k_1)$ are "Lipschitz continuous" with Lipschitz constant k_1, which explains the notation $\text{Lip}(k_1)$.

[12] As an aside, we note that the lemma has an immediate generalization to *metric spaces*. This may be useful in other studies.

in which μ is a finite signed Borel measure (Royden 1968, pp. 310, 302).[13] We have $|g(x)| \leq V\|x\|$, where $\|\cdot\|$ is the usual max norm and V denotes the total variation $|\mu|(I)$ of μ on I. The case in which $V = 0$ is trivial, so assume $V \neq 0$.

Let $\delta > 0$ and $x_1, \ldots, x_r$ in C be given. Using the continuity of the x_i, write I as the union of pairwise disjoint n-dimensional intervals[14] $I_1, \ldots, I_p$ containing points $\alpha_1, \ldots, \alpha_p$, respectively, such that

$$|x_i(\alpha) - x_i(\alpha_j)| \leq \delta/(2V)$$

for $\alpha \in I_j$ $(1 \leq j \leq p, 1 \leq i \leq r)$. Define $h: C \to \mathbb{R}$ by

$$hx = \sum_j x(\alpha_j) \int_{I_j} d\mu$$

Then

$$|hx| \leq \sum_j \|x\| \cdot |\mu|(I_j) = V\|x\| \qquad x \in C$$

and

$$|gx_i - hx_i| = \left| \int_I x_i \, d\mu - \sum_j x_i(\alpha_j) \int_{I_j} d\mu \right| \leq \sum_j \int_{I_j} |x_i(\alpha) - x_i(\alpha_j)| d|\mu|$$

$$\leq \frac{\delta}{2V} \sum_j |\mu|(I_j) = \frac{\delta}{2} < \delta$$

for each i. Therefore, by the lemma, we have proved the first part of the theorem.

From the proof above (or by the equicontinuity of the elements of C), it is clear that the α_j can be chosen so that they are interior points of I. We assume that the α_j meet this condition in what follows. To complete the proof it suffices to show that for any $h: C \to \mathbb{R}$ given by

---

[13] The concept of measure is an extension of the concept of n-dimensional volume. A Borel measure on I is a measure defined on the Borel subsets of I such that it is finite for each compact set. The family of Borel sets is the smallest σ algebra (Royden 1968) containing the closed sets.

[14] An n-dimensional interval is a set of the form $\Delta_1 \times \cdots \times \Delta_n$ where each Δ_k is an interval in $\mathbb{R}$ that may be open, closed, or half open.

$$hx = \sum_j c_j x(\alpha_j)$$

there is a $q \in X$ such that

$$\left| h(x) - \int_I q(\alpha)x(\alpha)d\alpha \right| < \frac{\varepsilon}{2} \qquad x \in C$$

We assume below that $c_j \neq 0$ for some j.

Choose $\sigma > 0$ so that the sets

$$A_j := [\alpha_{j1} - \sigma/2, \alpha_{j1} + \sigma/2] \times \cdots \times [\alpha_{jn} - \sigma/2, \alpha_{jn} + \sigma/2]$$

($j = 1, \ldots, p$) are pairwise disjoint subsets of the interior of I and (using the equicontinuity of the members of C) so that

$$|x(\beta_1) - x(\beta_2)| < \varepsilon/(4c)$$

for $x \in C$ with $\max_k |\beta_{1k} - \beta_{2k}| \leq \sigma/2$, where $c = \Sigma_j |c_j|$. For each j define $s_j \colon I \to \mathbb{R}$ by $s_j(\alpha) = 1/\sigma^n$ for $\alpha \in A_j$ and $s_j(\alpha) = 0$ otherwise.

We have for any $x \in C$

$$\left| \int_I \sum_j c_j s_j(\alpha)x(\alpha)\,d\alpha - \sum_j c_j x(\alpha_j) \right| \leq \sum_j |c_j| \frac{1}{\sigma^n} \int_{A_j} |x(\alpha_j) - x(\alpha)|\,d\alpha$$

$$< \sum_j \frac{|c_j|\varepsilon}{4c} \leq \frac{\varepsilon}{4}$$

So, with $s = \Sigma_j c_j s_j$,

$$\left| \int_I s(\alpha)x(\alpha)\,d\alpha - \sum_j c_j x(\alpha_j) \right| < \frac{\varepsilon}{4} \qquad x \in C$$

We now construct a suitable continuous approximation to s. Let $\eta > 0$ be sufficiently small that the sets

$$B_j := (\alpha_{j1} - \sigma/2 - \eta, \alpha_{j1} + \sigma/2 + \eta) \times \cdots$$
$$\times (\alpha_{jn} - \sigma/2 - \eta, \alpha_{jn} + \sigma/2 + \eta)$$

$(j = 1, \ldots, p)$ are pairwise disjoint subsets of I. Choose $q(\alpha) = s(\alpha)$ for $\alpha \in \cup_j A_j$ and $q(\alpha) = 0$ for $\alpha \in I - \cup_j B_j$. Since $I - \cup_j B_j$ together with the A_j are pairwise disjoint closed subsets of $\mathbb{R}^n$, it is possible to complete the definition of q on I so that it is continuous and its values lie in a finite interval $[\gamma_1, \gamma_2]$ that does not depend on η [see Natanson (1960, p. 81)].[15] Having done that we see that

$$\int_I |q(\alpha) - s(\alpha)| \, d\alpha = \int_{\cup_j B_j} |q(\alpha) - s(\alpha)| \, d\alpha$$
$$= \int_{\cup_j (B_j - A_j)} |q(\alpha)| \, d\alpha$$

which can be made arbitrarily small by choosing η sufficiently small. Thus, since

$$\max_{x \in C} \max_{\alpha} |x(\alpha)| < \infty$$

it is clear that there is a continuous $q \colon I \to \mathbb{R}$ such that

$$\left| \int_I [q(\alpha) - s(\alpha)] x(\alpha) \, d\alpha \right| < \frac{\varepsilon}{4} \qquad x \in C$$

Thus,

$$\left| \int_I q(\alpha) x(\alpha) \, d\alpha - \sum_j c_j x(\alpha_j) \right| < \frac{\varepsilon}{2}$$

for $x \in C$, which completes the proof.

2.2.3 An Example of a Classifier

In this section we give a simple example of a classifier for classifying continuous signals of n variables. Application areas corresponding to $n = 1, 2,$ and 3 include automatic target identification and pattern recognition in two and three dimensions, respectively.

Let X be the space of continuous real-valued functions defined on $[0, 1]^n$ with $\|\cdot\|$ the usual sup norm, let κ and r be positive constants, and let $\text{Lip}(\kappa)$ denote the subset of X consisting of the elements x of

---

[15] We can take $\gamma_1 = \sigma^{-n} \min\{0, c_1, \ldots, c_p\}$ and $\gamma_2 = \sigma^{-n} \max\{0, c_1, \ldots, c_p\}$.

X that satisfy $|x(a) - x(b)| \leq \kappa |a - b|$ for all a and b, where $|a - b|$ stands for the Euclidean norm of $a - b$. Let $x_{(1)}, \ldots, x_{(m)}$ be distinct elements of $\mathrm{Lip}(\kappa)$, and take $C_j = \{x \in \mathrm{Lip}(\kappa): \|x - x_{(j)}\| \leq r\}$ for each $j = 1, \ldots, m$.

Assume that $r < \frac{1}{2}\min_{i \neq j}\|x_{(i)} - x_{(j)}\|$ in which case the C_j are pairwise disjoint. Since each C_j is a closed bounded subset of X that is equicontinuous on $[0, 1]^n$, by a version of the Arzela–Ascoli theorem (Mukherjea and Pothoven, 1978, p. 76)[16] the C_j are compact. Using the corollary and Theorem 2 we see that classification of the signals in $\cup_j C_j$ is possible with the structure of Fig. 2.1 with the functionals y_j performing either a simple sampling and summing operation or a simple integration. Acceptable choices of h in Fig. 2.1 and of the u_j in Fig. 2.2 are described in Section 2.2.1.

The example described above can be generalized in several directions. For instance, the C_j need not be closed balls; they can be any pairwise disjoint collection of closed bounded subsets of X whose elements belong to $\mathrm{Lip}(\kappa)$. Notice that we have obtained what amounts to a precise mathematical statement of a classification result that goes significantly beyond what has previously been shown to be achievable using neural networks.[17]

Classification results along the lines described above can be given for many other spaces X using the various criteria for compactness available in the mathematics literature.

2.2.4 Approximations on $\mathbb{R}^n$ Using Sigmoids

Theorem 1 has applications other than to classification problems. For example, there is an interesting result often attributed to Cybenko (1989) that finite sums of the form

$$\sum_j \gamma_j s(w_j \cdot + \rho_j)$$

---

[16] The Arzela–Ascoli theorem is a tool theorem that provides conditions for compactness. In this theorem, a central role is played by an equicontinuity condition. That condition is met in the case considered above because the elements of each C_j belong to $\mathrm{Lip}(\kappa)$.

[17] The problem of classifying continuous-time signals is a familiar problem in the neural networks field where the primary focus has been on sampling the signals in order to use results concerning the approximation of real-valued functions of a finite number of variables. As is well known, this sampling introduces errors that typically are not addressed analytically, and conclusive statements that describe the families of signals that can be classified are typically not given.

can uniformly approximate continuous $\mathbb{R}$-valued functions on any compact subset of $\mathbb{R}^n$. Here the γ_j and ρ_j are real constants, the elements of $\mathbb{R}^n$ are viewed as column n vectors, each w_j is a real row n vector, and $s: \mathbb{R} \to \mathbb{R}$ is continuous and satisfies $\lim_{\beta \to \infty} s(\beta) = 1$ and $\lim_{\beta \to -\infty} s(\beta) = 0$. The (i) $\Rightarrow$ (ii) part of Theorem 1 provides an immediate proof of Cybenko's result,[18] given that it is true for $n = 1$, because in this case we can choose U to be the set of sigmoidal functions indicated in an earlier footnote and can take Y to be the set of linear functionals ϕ in $\mathbb{R}^n$ (with any norm); these linear functionals have the representation $\phi(x) = wx$ where w is a real row n vector.

2.3 MYOPIC MAPS, NEURAL NETWORK APPROXIMATIONS, AND VOLTERRA SERIES

The most important single result in the theory of linear systems is that under conditions that are ordinarily satisfied, a linear system's output is represented by a convolution operator acting on the system's input. This result provides understanding concerning the behavior of linear systems as well as an analytical basis for their design. There is no such simple result for nonlinear systems, but certain classes of nonlinear systems possess Volterra series (or Volterra-like series) representations in which the output is given by a familiar infinite sum of iterated integral operators operating on the system's input. The pressing need to have available representations for nonlinear systems, as well as to understand the limitations and advantages of such representations, is illustrated by fact that Volterra series representations have been discussed and studied in applications prior to the appearance of any results that established their existence, and even recent publications often cite Volterra (1959) in seeking to justify the use of a Volterra series approach even though no justification is given (or intended) there.[19]

---

[18] For other approximation results of this kind see, for example, Hornik (1991).

[19] The most pertinent material in Volterra's path-breaking (Volterra 1959) is Fréchet's Weierstrass-like result in Volterra (1959, p. 20). While interesting and important, it has significant limitations with regard to Volterra series representations of input–output maps in that, for example, it concerns approximations rather than series expansions in the usual sense. An analogy is that while any continuous real function f on the interval $[0, 1]$ can be approximated uniformly arbitrarily well by a polynomial, this does not mean that every such f has a power series representation. Another limitation is that Fréchet's result directly concerns functionals rather than mappings

A Volterra series representation, when it exists, is a very special representation in that, under weak assumptions, the terms in the infinite sum are unique. This uniqueness, at least in principle, often facilitates the determination of the terms in the series. But just as there are many real-valued functions on the interval [0, 1] that do not have a power series expansion, there are many systems whose input–output maps do not have a Volterra series representation. However, the failure of a system to have an exact representation of a certain form does not rule out the possibility that a useful approximate representation may exist. This is the main reason that approximations to input–output maps are of interest.

The general problem of approximating input–output maps arises also in the neural networks field where much progress has been made in recent years. Initially attention was focused on the problem of approximating real-valued maps of a finite number of real variables. For example, as mentioned in Section 2.2.4, Cybenko (1989) proved that any real-valued continuous map defined on a compact subset of $\mathbb{R}^n$ (n an arbitrary positive integer) can be uniformly approximated arbitrarily well using a neural network with a single hidden layer using any continuous sigmoidal activation function. And subsequently Mhaskar and Micchelli (1992), and also Chen and Chen (1995), showed that even any nonpolynomial continuous function would do. There are also general studies of approximation in $L_p(\mathbb{R}^n)$, $1 \leq p < \infty$ (and on compact subsets of $\mathbb{R}^n$) using radial basis functions as well as elliptical basis functions. It is known (Park and Sandberg 1993), for example, that arbitrarily good approximation of a general $f \in L_1(\mathbb{R}^n)$ is possible using uniform smoothing factors and radial basis functions generated in a certain natural way from a single g in $L_1(\mathbb{R}^n)$ if and only if g has a nonzero integral.

The material just described is concerned with the approximation of systems without dynamics. At about the same time there began (Sandberg 1991) a corresponding study of the network (e.g., neural network) approximation of functionals and approximately finite-

---

from one function space to another, and the domain of Fréchet's functionals is a set of functions defined on an interval that is finite. In contrast, in most circuits and systems studies involving Volterra series, inputs are defined on $\mathbb{R}$ or $\mathbb{R}_+$ (or on the discrete-time analogs of these sets). For influential early material concerning Volterra series and applications, see Schetzen (1980). Additional background material, references, and a description of relatively recent results concerning the existence and convergence of Volterra series can be found in Sandberg (1984).

memory maps. It was shown that large classes of approximately finite-memory maps can be uniformly approximated arbitrarily well by the maps of certain simple nonlinear structures using, for example, sigmoidal nonlinearities or radial basis functions.[20] This is of interest in connection with, for example, the general problem of establishing a comprehensive analytical basis for the identification of dynamic systems.[21] The approximately finite-memory approach in Sandberg (1991) is different from, but is related to, the fading-memory approach in Boyd and Chua (1985) where it is proved that certain scalar single-variable causal fading-memory systems with inputs and outputs defined on $\mathbb{R}$ can be approximated by a finite Volterra series.

The study in Sandberg (1991) addresses noncausal as well as causal systems and also systems in which inputs and outputs are functions of several variables. In Sandberg and Xu (1997a) strong corresponding results are given within the framework of an extension of the fading-memory approach. A key idea in Sandberg and Xu (1997a) is a substantial generalization of the proposition in Boyd and Chua (1985) to the effect that a certain set of continuous functions defined on $\mathbb{R}$ is compact in a weighted-norm space. In Sandberg and Xu (1997a) attention is restricted to the case of inputs and outputs defined on the infinite m-dimensional interval $(-\infty, \infty)^m$ where m is an arbitrary positive integer. Here we consider the important "discrete-space" case in which inputs and outputs are defined on $(\ldots, -1, 0, 1, \ldots)^m$. We use the term *myopic* to describe the maps we study because the term *fading-memory* is a misnomer when applied to noncausal systems, in that noncausal systems may anticipate as well as remember. In our setting, a shift-invariant causal myopic map has what we call *uniform fading memory*.

---

[20] It was later found (Sandberg 1993b) [see also Sandberg (1992b)] that the approximately finite-memory condition is met by the members of a certain familiar class of stable continuous-time systems.

[21] As mentioned in Section 2.2, it was also observed that any continuous real functional on a compact subset of a real normed linear space can be uniformly approximated arbitrarily well using only a feedforward neural network with a linear-functional input layer and one hidden nonlinear (e.g., sigmoidal) layer. This is a kind of general extension of an idea due to Wiener concerning the approximation of input–output maps using a structure consisting of a bank of linear maps followed by a memoryless map with several inputs and a single output [see e.g., Schetzen (1980, pp. 380–382)]. For further results along these lines, see Sandberg and Xu (1996). And for related work, see Chen and Chen (1993).

In Section 2.3.1 we consider multidimensional shift-invariant input–output maps with vector-valued inputs drawn from a certain large set. Our main result is a theorem showing that if these maps satisfy a certain "myopicity" condition, which is very often met, then they can be uniformly approximated arbitrarily well using a structure consisting of a linear preprocessing state followed by a memoryless nonlinear network. Such structures were first considered in an important but very special context by Wiener (Schetzen 1980, pp. 380–382). We consider causal as well as noncausal maps. As we have said, approximations for noncausal maps for which inputs and outputs are functions of more than one variable, with each variable taking values in $(\ldots, -1, 0, 1, \ldots)$, are of current interest in connection with, for example, image processing.

Section 2.3.1 begins with a description of notation and key definitions. Our main result concerning myopic maps is given in the next section. This is followed by a section on comments and one on the specialization of our result to generalized discrete-space finite Volterra series approximations.

2.3.1   Approximation of Myopic Maps

Preliminaries   Throughout Section 2.3, $\mathbb{R}$ is the set of reals, $\mathbf{Z}$ is the set of all integers, and $\mathbb{N}$ is the set of positive integers. Let n and m in $\mathbb{N}$ be arbitrary. $\|\cdot\|$ and $<\cdot, \cdot>$ are the Euclidean norm and inner product on $\mathbb{R}^n$, respectively, and for each $\alpha := (\alpha_1, \ldots, \alpha_m) \in \mathbf{Z}^m$, $|\alpha|$ stands for $\max_i |\alpha_i|$. With $\mathbf{Z}_- = \{\ldots, -1, 0\}$, $\mathbf{Z}_-^m$ denotes $(\mathbf{Z}_-)^m$.

For any positive integer n_0, let $C(\mathbb{R}^{n_0}, \mathbb{R})$ denote the set of continuous maps from $\mathbb{R}^{n_0}$ to $\mathbb{R}$, and let D_{n_0} stand for any subset of $C(\mathbb{R}^{n_0}, \mathbb{R})$ that is dense on compact sets, in the usual sense that given $\varepsilon > 0$ and $f \in C(\mathbb{R}^{n_0}, \mathbb{R})$, as well as a compact $V \subset \mathbb{R}^{n_0}$, there is a $q \in D_{n_0}$ such that $|f(v) - q(v)| < \varepsilon$ for $v \in V$. The D_{n_0} can be chosen in many different ways, and may involve, for example, radial basis functions, polynomial functions, piecewise linear functions, sigmoids, or combinations of these functions.

Let w be an $\mathbb{R}$-valued function defined on $\mathbf{Z}^m$ such that $w(\alpha) \neq 0$ for all α and $\lim_{|\alpha| \to \infty} w(\alpha) = 0$. With $X(\mathbf{Z}^m, \mathbb{R}^n)$ the set of all $\mathbb{R}^n$-valued maps on $\mathbf{Z}^m$, denote by X_w the normed linear space given by

$$X_w = \left\{ x \in X(\mathbf{Z}^m, \mathbb{R}^n) : \sup_{\alpha \in \mathbf{Z}^m} \|w(\alpha)x(\alpha)\| < \infty \right\}$$

with the norm

$$\|x\|_w = \sup_{\alpha \in \mathbf{Z}^m} \|w(\alpha)x(\alpha)\|$$

Later we will use the fact that X_w is complete (see Appendix A).

Now let $\chi(\mathbf{Z}^m, \mathbb{R}^n)$ be the set of all bounded functions contained in $X(\mathbf{Z}^m, \mathbb{R}^n)$, and let S be a nonempty subset of $\chi(\mathbf{Z}^m, \mathbb{R}^n)$. For each $\beta \in \mathbf{Z}^m$, define T_β on S by

$$(T_\beta x)(\alpha) = x(\alpha - \beta) \qquad \alpha \in \mathbf{Z}^m$$

The set S is said to be *closed under translation* if $T_\beta S = S$ for each $\beta \in \mathbf{Z}^m$.[22] Let G map S to the set of $\mathbb{R}$-valued functions on $\mathbf{Z}^m$. Such a G is *shift invariant* if S is closed under translation and

$$(Gx)(\alpha - \beta) = (GT_\beta x)(\alpha) \qquad \alpha \in \mathbf{Z}^m$$

for each $\beta \in \mathbf{Z}^m$ and $x \in S$. The map G is *causal* if for each $\beta \in \mathbf{Z}^m$ we have

$$(Gx)(\beta) = (Gy)(\beta)$$

whenever x and y belong to S and $x(\alpha) = y(\alpha)$ for $\alpha_j \leq \beta_j \forall_j$.

We assume throughout Section 2.3 that G is shift invariant. We say that G is *myopic* on S with respect to w if given an $\varepsilon > 0$ there is a $\delta > 0$ with the property that x and y in S and

$$\sup_{\alpha \in \mathbf{Z}^m} \|w(\alpha)[x(\alpha) - y(\alpha)]\| < \delta \Rightarrow |(Gx)(0) - (Gy)(0)| < \varepsilon \qquad (2.4)$$

Thus, and roughly speaking, G is myopic if the value of $(Gx)(\alpha)$ is always relatively independent of the values of x at points remote from α.

In our theorem in the next section we refer to certain sets $G(w)$ and $G_-(w)$. These sets concern sums of the form

---

[22] With regard to the choice of the words *closed under translation*, an equivalent definition is $T_\beta S \subseteq S$ for all β. To see this, observe that if $T_\beta S \subseteq S$ for all β and $T_{\beta_0} S \subset S$ for some β_0, then $T_{\beta_0} A \subset S$ for any $A \subseteq S$, which leads to the contradiction that $T_{\beta_0} T_{-\beta_0} S \subset S$.

$$\sum_{\beta \in D} \langle g(\beta), x(\beta) \rangle \qquad (2.5)$$

in which $x \in \chi(\mathbf{Z}^m, \mathbb{R}^n)$ and $D = \mathbf{Z}^m$ or $\mathbf{Z}_-^m$. Such sums are well defined in the sense of absolute summability (see Appendix B) for any g from D to $\mathbb{R}^n$ such that

$$\sum_{\beta \in D} \left\| w(\beta)^{-1} g(\beta) \right\| < \infty \qquad (2.6)$$

[because (2.6) implies the absolute summability of g]. Let $D = \mathbf{Z}^m$. By $G(w)$ we mean any set of functions g from D to $\mathbb{R}^n$ such that (2.6) is met for each g, and for each nonzero $x \in \chi(\mathbf{Z}^m, \mathbb{R}^n)$ there corresponds a g for which (2.5) is nonzero. Similarly, with $D = \mathbf{Z}_-^m$, the set $G_-(w)$ is any set of functions g from D to $\mathbb{R}^n$ such that (2.6) is met for each g, and for each $x \in \chi(\mathbf{Z}^m, \mathbb{R}^n)$ whose restriction to D is nonzero there is a g for which (2.5) is nonzero. The sets $G(w)$ and $G_-(w)$ can be chosen in many ways (see Appendix C). Finally, let Q be the map from S to $X(\mathbf{Z}^m, \mathbb{R}^n)$ defined by $(Qs)(\alpha) = s(\beta)$ for each s and α, where $\beta_j = -|\alpha_j|$ for all j.

Our Main Result   Our main result below concerning the approximation of myopic maps gives, in a certain setting, a necessary and sufficient condition for the uniform approximation of shift-invariant maps with vector-valued inputs of a finite number of variables.[23] In stating this result, we use the fact that sums of the form

$$\sum_{\beta \in \mathbf{Z}^m} \langle p(\beta - \alpha), x(\beta) \rangle$$

and

$$\sum_{\beta \in (-\infty, \alpha]} \langle q(\beta - \alpha), x(\beta) \rangle$$

are well defined and finite for each $\alpha \in \mathbf{Z}^m$ when x is an element of $\chi(\mathbf{Z}^m, \mathbb{R}^n)$, $p \in G(w)$, and $q \in G_-(w)$. This follows from the observa-

---

[23] The result was obtained jointly by this writer and his one-time graduate student L. Xu.

tion that by (2.6) both p and q are absolutely summable on their respective domains (see Appendix B).

Theorem 3. Assume that S is uniformly bounded.[24] (Recall that G is a shift-invariant map from S to the set of $\mathbb{R}$-valued functions defined on $\mathbf{Z}^m$.) Then the following two statements are equivalent.

    (i) G is myopic on S with respect to w.

   (ii) For each $\varepsilon > 0$, there are an $n_0 \in \mathbb{N}$, elements $g_1, \ldots, g_{n_0}$ of $G(w)$, and an $N \in D_{n_0}$ such that

$$\|(Gx)(\alpha) - N[(Lx)(\alpha)]\| < \varepsilon \qquad \alpha \in \mathbf{Z}^m \qquad (2.7)$$

for all $x \in S$, where L is given by

$$(Lx)_j(\alpha) = \sum_{\beta \in \mathbf{Z}^m} \langle h_j(\alpha - \beta), x(\beta) \rangle \qquad (2.8)$$

with $h_j(\beta) = g_j(-\beta)$ for all β and j.

Moreover, if G is causal and $QS \subseteq S$, then (ii) can be replaced with:

  (ii′) For each $\varepsilon > 0$, there are an $n_0 \in \mathbb{N}$, elements $g_1, \ldots, g_{n_0}$ of $G_-(w)$, and an $N \in D_{n_0}$ such that $\|(Gx)(\alpha) - N[(Lx)(\alpha)]\| < \varepsilon$ for all $\alpha \in \mathbf{Z}^m$ and $x \in S$, where L is given by

$$(Lx)_j(\alpha) = \sum_{\beta \in (-\infty, \alpha]} \langle h_j(\alpha - \beta), x(\beta) \rangle \qquad (2.9)$$

with $h_j(\beta) = g_j(-\beta)$ for all β and j, and where $(-\infty, \alpha]$ means $\{\ldots, (\alpha_1 - 1), \alpha_1\} \times \ldots \times \{\ldots, (\alpha_m - 1), \alpha_m\}$.

Proof. (i) $\Rightarrow$ (ii): Assume that (i) holds. Recall that $w(\alpha) \to 0$ as $|\alpha| \to \infty$ and that S, which is uniformly bounded, is a subset of the complete space X_w. By the characterization in Liusternik and Sobolev (1961, p. 136) of relative compactness of subsets of a general Banach space[25]

---

[24] This means that there is a positive constant c for which $\|x(\alpha)\| \le c$ for all $x \in S$ and all α.

[25] Recall that by a *Banach space* is meant a normed linear space that is complete in the sense that every Cauchy sequence in the space converges to an element of the space.

with a basis, it is not difficult to show that S is a relatively compact subset of X_w.[26] Since G is myopic with respect to w, the functional $(G\cdot)(0)$ is uniformly continuous on S with respect to $\|\cdot\|_w$. We will use the following lemma proved in Appendix D.

Lemma 2. Assume that U is a nonempty uniformly bounded subset of $X(\mathbf{Z}^m, \mathbb{R}^n)$ that is relatively compact in X_w. Let $F: U \to \mathbb{R}$ be uniformly continuous with respect to $\|\cdot\|_w$. Then given $\varepsilon > 0$, there are an $n_0 \in \mathbb{N}$, elements $g_1, \ldots, g_{n_0}$ of $\mathcal{G}(w)$, and an $N \in D_{n_0}$ such that

$$|F(u) - N[y_1(u), \ldots, y_{n_0}(u)]| < \varepsilon \qquad u \in U$$

where each y_j is defined on U by

$$y_j(u) = \sum_{\beta \in \mathbf{Z}^m} \langle g_j(\beta), u(\beta) \rangle$$

Continuing with the proof, let $\varepsilon > 0$, $\alpha \in \mathbf{Z}^m$, and $x \in S$ be given. By Lemma 2 with $U = S$, $F = (G\cdot)(0)$, and $u = T_{-\alpha}x$, we have

$$|G(T_{-\alpha}x)(0) - N[y_1(T_{-\alpha}x), \ldots, y_{n_0}(T_{-\alpha}x)]| < \varepsilon$$

with N, n_0, and the g_j as described in Lemma 2. By the shift invariance of G, $(Gx)(\alpha) = (GT_{-\alpha}x)(0)$. And since $T_{-\alpha}x(\beta) = x(\alpha + \beta)$ for all β, we have

$$y_j(T_{-\alpha}x) = \sum_{\beta \in \mathbf{Z}^m} \langle g_j(\beta), x(\alpha + \beta) \rangle = \sum_{\beta \in \mathbf{Z}^m} \langle g_j(\beta - \alpha), x(\beta) \rangle$$

for each j. Thus we have (ii).

(i) $\Rightarrow$ (ii'): Assume that $QS \subseteq S$ and that G is causal as well as myopic on S with respect to w. Here we use the following variant of Lemma 2 proved in Appendix E.

---

[26] Let $e_\alpha^j: \mathbf{Z}^m \to \mathbb{R}^n$ with $\alpha \in \mathbf{Z}^m$ and $j \in \{1, \ldots, n\}$ be defined by the condition that $e_\alpha^j(\beta) = 0$ for $\beta \neq \alpha$ and $e_\alpha^j(\beta)$ for $\beta = \alpha$ is the element of $\mathbb{R}^n$ with jth component unity and all other components zero. Then $\{e_\alpha^j\}$ is clearly a basis for X_w. The result in Liusternik and Sobolev (1961, p. 136) is applicable because $\|e_\alpha^j\|_w \to 0$ as $|\alpha| \to \infty$ uniformly in j.

Lemma 3. Let U be a nonempty uniformly bounded subset of $X(\mathbf{Z}^m,$ $\mathbb{R}^n)$ that contains QU. Assume also that U is relatively compact in X_w, and let $F: U \to \mathbb{R}$ be uniformly continuous with respect to $\|\cdot\|_w$. Then given $\varepsilon > 0$, there are an $n_0 \in \mathbb{N}$, elements $g_1, \ldots, g_{n_0}$ of $\mathcal{G}_-(w)$, and an $N \in D_{n_0}$ such that

$$|F(u) - N[y_1(u), \ldots, y_{n_0}(u)]| < \varepsilon \qquad u \in QU$$

where each y_j is defined on QU by

$$y_j(u) = \sum_{\beta \in \mathbf{Z}^m} \langle g_j(\beta), u(\beta) \rangle$$

Now let $x \in S$ and $\alpha \in \mathbf{Z}^m$ be given and, referring to Lemma 3, take $U = S$, $F = (G\cdot)(0)$, and $u = QT_{-\alpha}x$. Here for each j, $y_j(QT_{-\alpha}x) = y_j(T_{-\alpha}x)$ and

$$y_j(T_{-\alpha}x) = \sum_{\beta \in \mathbf{Z}^m} \langle g_j(\beta), x(\alpha + \beta) \rangle = \sum_{\beta \in (-\infty, \alpha]} \langle g_j(\beta - \alpha), x(\beta) \rangle$$

In addition, by the causality and shift invariance of G,

$$(GQT_{-\alpha}x)(0) = (GT_{-\alpha}x)(0) = (Gx)(\alpha)$$

This gives (ii').

(ii) $\Rightarrow$ (i): Assume that (ii) holds. Fix $\varepsilon > 0$, $u_1 \in S$, and $u_2 \in S$. By (ii) choose n_0, elements $g_1, \ldots, g_{n_0}$ of $\mathcal{G}(w)$, and an $N \in D_{n_0}$, all independent of u_1 and u_2, such that $|(Gu_1)(0) - N[(Lu_1)(0)]| < \varepsilon/3$ and $|(Gu_2)(0) - N[(Lu_2)(0)]| < \varepsilon/3$.

By the triangle inequality in $\mathbb{R}$,

$$|(Gu_1)(0) - (Gu_2)(0)| < 2\varepsilon/3 + |N[(Lu_1)(0)] - N[(Lu_2)(0)]|$$

Since N is continuous and (by the absolute summability of the g_j) $(LS)(0)$ is bounded, N is uniformly continuous on $(LS)(0)$. Thus, there is a $\delta_0 > 0$, independent of u_1 and u_2, such that

$$\|(Lu_1)(0) - (Lu_2)(0)\| < \delta_0 \Rightarrow |N[(Lu_1)(0)] - N[(Lu_2)(0)]| < \varepsilon/3$$

It thus suffices to show that there exists a $\delta > 0$, independent of u_1 and u_2, such that

$$\sup_{\alpha \in \mathbf{Z}^m} \|w(\alpha)[u_1(\alpha) - u_2(\alpha)]\| < \delta \Rightarrow \|(Lu_1)(0) - (Lu_2)(0)\| < \delta_0$$

For $j = 1, \ldots, n_0$ we have

$$\begin{aligned}
|(Lu_1)_j(0) - (Lu_2)_j(0)| &\le \sum_{\alpha \in \mathbf{Z}^m} |\langle g_j(\alpha), u_1(\alpha) - u_2(\alpha)\rangle| \\
&\le \sum_{\alpha \in \mathbf{Z}^m} \left\|w(\alpha)^{-1} g_j(\alpha)\right\| \cdot \|w(\alpha)[u_1(\alpha) - u_2(\alpha)]\| \\
&\le \sup_{\alpha \in \mathbf{Z}^m} \|w(\alpha)[u_1(\alpha) - u_2(\alpha)]\| \cdot \sum_{\alpha \in \mathbf{Z}^m} \left\|w(\alpha)^{-1} g_j(\alpha)\right\|
\end{aligned}$$

from which it is clear that we can choose δ as required.[27] (ii$'$) $\Rightarrow$ (i) can be established similarly. This completes the proof of the theorem.

2.3.2 Comments

The approximating structure of Theorem 3 is illustrated in Fig. 2.3. Our proof shows that (i) implies (ii) and (ii$'$) even without the hypothesis that the elements of D_{n_0} are continuous. Also, (ii) implies (i), and (ii$'$) implies (i), without the assumption that S is closed under translation.

The map Q in the theorem can be replaced with any map from S to $X(\mathbf{Z}^m, \mathbb{R}^n)$ such that $(Qx)(\alpha) = x(\alpha)$ for α in $\mathbf{Z}^m_-$ and $x \in S$, and for distinct elements x and y of QS the restriction $(x - y)|_{\mathbf{Z}^m_-}$ of $(x - y)$ to $\mathbf{Z}^m_-$ is not zero. For example, Q can be taken to be defined by

$$(Qx)(\alpha) = x(\hat{\alpha}) \qquad \alpha \in \mathbf{Z}^m \tag{2.10}$$

where $\hat{\alpha}$ is the unique point in $\mathbf{Z}^m_-$ that minimizes $|\alpha - \hat{\alpha}|$.[28]

The following condition is closely related to the concept of myopic maps: Given an $\varepsilon > 0$, there is a $\delta > 0$ such that x and y in S and

---

[27] A related but different way to argue that (ii) $\Rightarrow$ (i) is to observe that under (ii) the real-valued function $(G \cdot)(0)$ is uniformly continuous on S with respect to the norm in X_w, since $(G \cdot)(0)$ can be uniformly approximated arbitrarily well over S by uniformly continuous functions.

[28] $\hat{\alpha}$ is defined by the simple condition that each of its elements belongs to $\{\ldots, -1, 0\}$ and is as close as possible to the corresponding element of α.

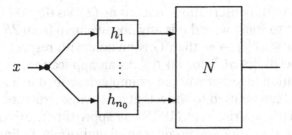

Figure 2.3 Approximation structure.

$$\sup_{\alpha \in \mathbf{Z}^m} \|\mu(\alpha)[x(\alpha) - y(\alpha)]\| < \delta \Rightarrow |(Gx)(0) - (Gy)(0)| < \varepsilon \qquad (2.11)$$

(note the $\mathbf{Z}_-^m$ instead of $\mathbf{Z}^m$), where μ is an $\mathbb{R}$-valued function on $\mathbf{Z}^m$ that satisfies $\lim_{|\alpha| \to \infty} \mu(\alpha) = 0$ and never vanishes. Appendix F is a brief study of implications concerning myopic maps, causality, and condition (2.11). In particular, it is shown there that, under the hypotheses of Theorem 3 and assuming $QS \subseteq S$, condition (2.11) is met for some μ if and only if G is causal and myopic on S with respect to some w. The uniform boundedness hypothesis of Theorem 3 concerning S, as well as the requirement that $QS \subseteq S$, are often satisfied. For example, they are satisfied for S the subset of $X(\mathbf{Z}^m, \mathbb{R}^n)$ consisting of all functions that are uniformly bounded with some bound c. This case with $m = n = 1$ is considered in Boyd and Chua (1985). And condition (2.11) is closely related to the *fading-memory* concept described there for causal G's. In this context condition (2.11) is what might be called a *uniform fading-memory* condition.

With S as described in Theorem 3, the condition that G is myopic on S with respect to w is equivalent to the seemingly weaker condition that given $\varepsilon > 0$ and $x \in S$ there is a $\delta > 0$ such that $|(Gx)(0) - (Gy)(0)| < \varepsilon$ for $y \in S$ with $\sup_{\alpha \in \mathbf{Z}^m} \|w(\alpha)[x(\alpha) - y(\alpha)]\| < \delta$. This is true because any continuous $\mathbb{R}$-valued function defined on a compact set is uniformly continuous, and the extension described in Appendix D of the functional F of Lemma 2 is defined on a compact set.[29]

---

[29] Similarly, it can be shown that, under the uniform boundedness hypothesis of Theorem 3, condition (2.11) is equivalent to the condition that given $\varepsilon > 0$ and $x \in S$ there is a $\delta > 0$ such that $|(Gx)(0) - (Gy)(0)| < \varepsilon$ for $y \in S$ with $\sup_{\alpha \in \mathbf{Z}_-^m} \|\mu(\alpha)[x(\alpha) - y(\alpha)]\| < \delta$. This for $m = n = 1$ is (essentially) the discrete-time fading-memory condition that plays a role in Boyd and Chua (1985). It can also be shown that, under the hypotheses of Theorem 3, (ii') holds if G is causal and condition (2.11) is met with μ the restriction to $\mathbb{R}_-^m$ of w. (The condition that $QS \subseteq S$ is not needed.)

A simple useful observation concerning (2.6) is this: If G is myopic with respect to some w, and q is any positive map from $\mathbf{Z}^m$ to $\mathbb{R}$ such that $q(\beta) \to \infty$ as $|\beta| \to \infty$, then G is myopic with respect to some w for which $|w(\beta)^{-1}| \le q(\beta)$ for all β.[30] As an application, we note that this observation together with an example described in a footnote in Appendix C can be used to show that a network structure related to one studied in De Vries et al. (1991) can approximate arbitrarily well any discrete-time time-invariant causal uniformly fading-memory input–output map for which the conditions on inputs of Theorem 3 are met and $m = n = 1$. For related results see, for example, Sandberg and Xu (1996, 1997b).

The class of maps G addressed by our theorem that are myopic with respect to some w is very large. In fact, a natural question to ask is whether there are G's that are not myopic. It turns out that one answer to this question is closely related to the theory of representations of linear G's and to an interesting pertinent oversight in the literature. This is discussed in Appendix G, where an example is given of a linear G that is myopic for no w.

We do not consider in this chapter the problem of actually determining the elements of the structure in Fig. 2.3. What we have done is to show that in an important setting it is possible to choose acceptable elements. And, of course, it is ordinarily useful to know that certain elements exist before one attempts to determine them by analytical, adaptive, or experimental techniques. Another illustration of our viewpoint is the following. In the area of signal processing there is considerable current interest in exploring nonlinear filtering techniques to overcome some of the basic limitations of linear filtering. Our theorem shows that the members of a certain class of nonlinear filters can be represented (as accurately as one wishes) by the special structure of Fig. 2.3. These are what might be called the *myopic-map filters*, and they form an interesting very large class of filters. In the next section we observe that the structure can provide Volterra series approximations.

2.3.3 Finite Generalized Volterra Series Approximations

If the D_{n_0} are chosen to be certain sets of polynomial functions, the approximating maps $N[L(\cdot)]$ of (ii) and (ii′) of Theorem 3 are gen-

---

[30] This is true because if G is myopic with respect to w, then it is myopic with respect to $|w|$, and thus with respect to w defined by $w(\beta) = \max\{|w(\beta)|, q(\beta)^{-1}\}$. A similar observation is made in Boyd and Chua (1985).

eralizations of discrete-time finite Volterra series. More specifically, suppose that the maps q of each D_{n_0} are such that $q(v)$ is a real polynomial in the components of v, and consider the $n = 1$ case. By writing products of sums as iterated sums, it is not difficult to see that $N[L(\alpha)]$ of (ii) has the form

$$k_0 + \sum_{j=1}^{p} \sum_{\beta_j \in \mathbf{Z}^m} \cdots \sum_{\beta_1 \in \mathbf{Z}^m} k_j(\alpha - \beta_1, \ldots, \alpha - \beta_j) x(\beta_1) \cdots x(\beta_j) \qquad (2.12)$$

in which k_0 is a constant, p is a positive integer, and each $k_j(\alpha - \beta_1, \ldots, \alpha - \beta_j)$ is a finite linear combination of products of the form $h_{i_j(1)}(\alpha - \beta_1) h_{i_j(2)}(\alpha - \beta_2) \ldots h_{i_j(j)}(\alpha - \beta_j)$, where the indices $i_j(1), \ldots, i_j(j)$ are drawn from $\{1, \ldots, n_0\}$ and the h's are as described in (ii). Similarly, in the case of (ii'), $N[L(\alpha)]$ has the form

$$k_0 + \sum_{j=1}^{p} \sum_{\beta_j \in (-\infty, \alpha]} \cdots \sum_{\beta_1 \in (-\infty, \alpha]} k_j(\alpha - \beta_1, \ldots, \alpha - \beta_j) x(\beta_1) \cdots x(\beta_j) \qquad (2.13)$$

where the k_j are formed from the h_j of (ii') in the way described above in connection with (ii). For each j, each of the two j-fold iterated sums in the summations over j of sums in (2.12) and (2.13) is a sum of products of sums over $\mathbf{Z}^m$ and $(-\infty, \alpha]$, respectively. Since in both cases the h_j are absolutely summable, by a result in Appendix B (and the boundedness of x) each sum in each product can be written as an m-fold iterated sum. Thus, each of the j-fold iterated sums can be simplified in that each can be written as a (jm)-fold iterated sum in which each summation is with respect to a scalar variable. For example, for $m = j = 2$,

$$\sum_{\beta_j \in \mathbf{Z}^m} \cdots \sum_{\beta_1 \in \mathbf{Z}^m} k_j(\alpha - \beta_1, \ldots, \alpha - \beta_j) x(\beta_1) \cdots x(\beta_j)$$

can be written in the form

$$\sum_{\gamma_4 \in \mathbf{Z}} \cdots \sum_{\gamma_1 \in \mathbf{Z}} \tilde{k}_2(\alpha_1 - \gamma_1, \alpha_2 - \gamma_2, \alpha_1 - \gamma_3, \alpha_2 - \gamma_4) x(\gamma_1, \gamma_2) x(\gamma_3, \gamma_4)$$

where $x(\gamma_1, \gamma_2) = x[(\gamma_1, \gamma_2)]$, $x(\gamma_3, \gamma_4) = x[(\gamma_3, \gamma_4)]$, and $\tilde{k}_2$ is related to k_2 in the obvious way.

Theorem 3 addresses only input–output maps that are shift invariant. Some thoughts on Volterra series approximations for shift-varying maps are described in Appendix H.

2.4   SEPARATION CONDITIONS AND APPROXIMATION OF DISCRETE-TIME AND DISCRETE-SPACE SYSTEMS

In Section 2.3, as well as in Sandberg (1991), Boyd and Chua (1993), and Sandberg and Xu (1997a) and other studies, attention centers around properties of nonlinear approximation structures of the type indicated in Fig. 2.3 in which the box labeled N is a memoryless nonlinear system and the h_j denote linear maps. This is a structure consisting of a linear preprocessing stage followed by a memoryless nonlinear network. As mentioned earlier, such structures were first considered in an important but very special context by Wiener. Roughly speaking, the main result in Section 2.3 is that, with N containing sigmoids, or radial basis functions, etc., a given shift-invariant input–output map G can be uniformly approximated over a certain set S of inputs if and only if G is myopic, assuming that the linear maps represented by the h_j satisfy certain conditions. Corresponding results for a different type of input sets are given in Sandberg and Xu (1997a, 1998).

In this section attention is focused on the h_j. We first consider causal time-invariant input–output maps G that take a set S of bounded vector-valued functions into a set of real-valued functions, and we give conditions on the h_j under which these G's can be uniformly approximated arbitrarily well using the structure shown in Fig. 2.3. In our results certain separation conditions, of the kind associated with the Stone–Weierstrass theorem, play a prominent role. Here they emerge as criteria for approximation and not just sufficient conditions under which an approximation exists. As an application of the results, we show that system maps of the type addressed can be uniformly approximated arbitrarily well by doubly finite Volterra series approximants[31] *if and only if* these maps have approximately finite memory and satisfy certain continuity conditions. Corresponding results are

---

[31] By such an approximant we mean one of the form $\Sigma_{j=1}^{p} \Sigma_{n_j=0}^{n} \ldots \Sigma_{n_1=0}^{n} k_j(n - n_1, \ldots, n - n_j)s(n_1) \cdots s(n_j)$, in which p is finite and for each j we have $k_j(n_1, \ldots, n_j) = 0$ if one or more of the $n_1, \ldots, n_j$ exceed some number η (which implies that the approximant has finite memory).

then given for (not necessary causal) multivariable input–output maps. As already mentioned, such multivariable maps are of interest in connection with image processing.

Our results are given in the next section, which begins with some preliminaries. As we have said, in these results certain separation conditions, of the kind associated with the Stone–Weierstrass theorem, play a prominent role, but here they emerge as criteria for approximation and not just as sufficient conditions under which an approximation exists. In particular, a corollary of one of our main results in the next section is a theorem to the effect that universal approximation can be achieved using the structure of Fig. 2.3 *if and only if* the set H from which the h_j are drawn satisfies the separation condition that for each n: $(hu_1)(n) \neq (hu_2)(n)$ for some $h \in H$ whenever $u_1, u_2 \in S$ and $u_1(j) \neq u_2(j)$ for at least one $j \leq n$. This holds even if the elements of H are not linear.[32] Results concerned with multivariable input–output maps, and with multivariable doubly finite Volterra series approximations, are given in Section 2.4.4 and Appendix K.

There are differences in the settings of Sections 2.3 and 2.4. Here the definition of a myopic map is different (but the main idea is similar), and an important role is played by the related concept of maps that possess "approximately finite memory" in a setting in which inputs and outputs are defined on the nonnegative integers. These differences are not actually needed to obtain results of the kind described above [see Sandberg (1998c)], and are introduced to illustrate an additional perspective.

2.4.1 Approximation of Input–Output Maps

Preliminaries The *linear-ring operations* starting with a set of real numbers consist of the linear operations and multiplication. That is, these operations consist of ordinary addition, multiplication, and multiplication by real scalars. Let k be a positive integer. We say that a map $M: \mathbb{R}^k \to \mathbb{R}$ is a *linear-ring map* if Mv is generated from the components $v_1, \ldots, v_k$ of v by a finite number of linear-ring operations that do not depend on v. Let $L(\mathbb{R}^k, \mathbb{R})$ stand for the set of all

---

[32] For a related result in the context of "complete memories," see Stiles et al. (1997) [see also Sandberg and Xu (1996, Theorem 4) and Sandberg (1998c)].

linear-ring maps from $\mathbb{R}^k$ to $\mathbb{R}$. We view the elements of $\mathbb{R}^k$ as row vectors.

Let $C(\mathbb{R}^k,\mathbb{R})$ denote the set of continuous maps from $\mathbb{R}^k$ to $\mathbb{R}$, and let D_k in Section 2.4 stand for any subset of $C(\mathbb{R}^k,\mathbb{R})$ that is dense in $L(\mathbb{R}^k,\mathbb{R})$ on compact sets, in the sense that given $\varepsilon > 0$ and $f \in L$ $(\mathbb{R}^k,\mathbb{R})$, as well as a compact $K \subset \mathbb{R}^k$, there is a $g \in D_k$ such that $|f(v) - g(v)| < \varepsilon$ for $v \in K$. The D_k can be chosen in many different ways and may involve, for example, radial basis functions, polynomial functions, piecewise linear functions, sigmoids, or combinations of these functions.[33]

Let d be a positive integer, and let C stand for any bounded closed subset of $\mathbb{R}^d$ that contains the origin of $\mathbb{R}^d$. For example, C can be chosen to be $\{v \in \mathbb{R}^d : \|v\| \leq \gamma\}$ where $\|\cdot\|$ is any norm on $\mathbb{R}^d$ and γ is a positive constant. With $\mathbf{Z}_+ =: \{0, 1, \dots\}$, let S denote the family of all maps s from $\mathbf{Z}_+$ to C. The set S is our set of inputs.

For each α and β in $\mathbf{Z}_+$, let maps $W_{\beta,\alpha} : S \to S$ and $T_\beta : S \to S$ be defined by

$$(W_{\beta,\alpha}s)(n) = \begin{cases} s(n) & \beta - \alpha \leq n \leq \beta \\ 0 & \text{otherwise} \end{cases}$$

and

$$(T_\beta s)(n) = \begin{cases} 0 & n < \beta \\ s(n - \beta) & n \geq \beta \end{cases}$$

We say that a map M from S into the set of real-valued functions on $\mathbf{Z}_+$ is *time invariant* if for each $\beta \in \mathbf{Z}_+$ we have

$$(MT_\beta s)(n) = \begin{cases} 0 & n < \beta \\ (Ms)(n - \beta) & n \geq \beta \end{cases}$$

for all s. M is *causal* if $(Mu)(n) = (Mv)(n)$ whenever $n \in \mathbf{Z}_+$ and u and v satisfy $u(\alpha) = v(\alpha)$ for $\alpha \leq n$.

Throughout Section 2.4, G denotes a causal time-invariant map from S to the set of real-valued functions defined on $\mathbf{Z}_+$. We assume

---

[33] The term D_k is used also in Section 2.3.1 where the meaning is different. Here the conditions on the D_k are even less restrictive.

that G has *approximately finite memory* in the sense that given $\varepsilon > 0$ there is an $\alpha \in \mathbf{Z}_+$ such that

$$|(Gs)(n) - (GW_{n,\alpha}s)(n)| < \varepsilon \qquad n \in \mathbf{Z}_+$$

for $s \in S$.[34] {The concepts of maps that are myopic or have approximately finite memory, fading memory (Boyd and Chua 1985), or decaying memory (Sandberg 1983) are all different but are all related in that they are alternative ways of making precise, in different settings, the same general idea. There is also a history of the use of this idea in other areas and for purposes other than approximation [see, e.g., Coleman and Mizel (1966, 1968), Sandberg (1992b), and Shamma and Zhao (1993)]}.

For each $n \in \mathbf{Z}_+$, let c_n stand for $\{0, 1, \ldots n\}$, and let S_n denote the restriction of S to c_n. We view each S_n as a metric space with metric ρ_n defined by $\rho_n(x, y) = \max \{\|x(j) - y(j)\|: j \in c_n\}$. We shall use the fact that each S_n is compact.[35]

Let H be a family of time-invariant causal maps h from S to the set of $\mathbb{R}$-valued functions defined on $\mathbf{Z}_+$, and for each h and each n in $\mathbf{Z}_+$ define the functional $q(h,n,\cdot)$ on S_n by $q(h,n,u) = (hs)(n)$, where s is any element of S whose restriction to c_n is u. We assume that $q(h, n, \cdot)$ is continuous for each $h \in H$. We also assume that H is closed under the memory-limiting operation, in the sense that g defined on S by $(gs)(n) = (hW_{n,\alpha}s)(n)$ belongs to H whenever $h \in H$ and $\alpha \in \mathbf{Z}_+$.[36]

As an example, we can take H to be the set H_0 of all maps h for which

$$(hs)(n) = \phi\left[\sum_{j=0}^{n} s(j)a(n - j) \right] \qquad n \in \mathbf{Z}_+ \qquad (2.14)$$

where ϕ, which depends on h, is a continuous map from $\mathbb{R}$ into $\mathbb{R}$ with $\phi(0) = 0$, and a, which also depends on h, is real $d \times 1$ matrix valued with $a(j)$ the zero $d \times 1$ matrix for j sufficiently large. As another example, note that H can be taken to be any subset of H_0 that is closed under the memory-limiting operation.

---

[34] There is a slight difference here relative to the definition of approximately finite memory in Sandberg (1991) where α is required to be positive.

[35] S_n is the continuous image of a compact set in $\mathbb{R}^{(n+1)r}$.

[36] In this connection, it is not difficult to check that g defined above is causal and time invariant for $h \in H$ and $\alpha \in \mathbf{Z}_+$.

For each $n \in \mathbf{Z}_+$, let F_n denote the functional defined on S_n by $F_n u = (Gs)(n)$, where s is any element of S whose restriction to c_n is u. We shall use A.1 to denote the following condition:

For each $n \in \mathbf{Z}_+$ and each $(u_1, u_2) \in E_n$, there is an $h \in H$ such that $q(h, n, u_1) \neq q(h, n, u_2)$

in which

$$E_n = \{(u_1, u_2) \in S_n \times S_n : F_n u_1 \neq F_n u_2\}.$$

2.4.2 Approximation and Discrete-Time Systems

One of our main results in this subsection is the following.

Theorem 4. The following two statements are equivalent.

(i) For each $\varepsilon > 0$, there are an $\alpha \in \mathbf{Z}_+$, a positive integer k, elements $h_1, \ldots, h_k$ of H, and an $N \in D_k$ such that

$$|(Gs)(n) - N[(MW_{n,\alpha}s)(n)]| < \varepsilon \qquad n \in \mathbf{Z}_+$$

for all $s \in S$, where $(Ms)(n) = [(h_1 s)(n), \ldots, (h_k s)(n)]$.

(ii) Each F_n is continuous and A.1 is met.

Poof. We will use the following lemma.

Lemma 4. Let A be a compact topological space,[37] let f belong to the set C of all continuous real-valued functions on A, and let B be a subset of C. Suppose that there is an element a of A such that $f(a) = 0$ and $b(a) = 0$ for $b \in B$. Then statements 1 and 2 below are equivalent.

1. For each $\varepsilon > 0$ there are a positive integer k, elements $b_1, \ldots, b_k$ of B, and an $N \in D_k$ such that $|f(x) - N[B(x)]| < \varepsilon$ for $x \in A$, where $B(x) = [(b_1)(x), \ldots, (b_k)(x)]$.

---

[37] This lemma is stated in a more general way than is needed in this chapter, in that the word *topological* could have been replaced with the word *metric* (since the topological spaces of interest to us are metric spaces). The lemma is stated in its more general form mainly to make more clear its relation to material in Stone (1962).

2. For $(x_1, x_2) \in \{(x_1, x_2) \in A \times A: f(x_1) \neq f(x_2)\}$ there is a $b \in B$ such that $b(x_1) \neq b(x_2)$.

To prove the lemma we make use of the following result, which we will refer to as Stone's theorem.

Theorem 5. Let A be a compact topological space, let f belong to the set C of all continuous real-valued functions on A, and let B be a subset of C. Then statements 1 and 2 below are equivalent.

1. For each $\varepsilon > 0$ there are a positive integer k, elements $b_1, \ldots, b_k$ of B, and a linear-ring map R such that $\|f(x) - R[B(x)]\| < \varepsilon$ for $x \in A$, where $B(x) = [(b_1)(x), \ldots, (b_k)(x)]$.
2. For $x \in \{x \in A: f(x) \neq 0\}$ there is a $b \in B$ such that $b(x) \neq 0$, and for $(x_1, x_2) \in \{(x_1, x_2) \in A \times A: f(x_1) \neq f(x_2)\}$ there is a $b \in B$ such that $b(x_1) \neq b(x_2)$.

Stone's theorem is a restatement of part of Theorem 5 of Stone (1962). Statement 1 is equivalent to a statement in Stone (1962, Theorem 5) concerning the family $\mathcal{U}(B)$ of functions generated from B by the linear-ring operations and uniform passage to the limit, and statement 2 is the contrapositive of the condition given in Stone (1962, Theorem 5) that f satisfy every linear relation $b(x) = 0$ or $b(x) = b(y)$ that is satisfied by all functions in B.[38]

Continuing with the proof of the lemma, let statement 2 of the lemma be satisfied and let $x \in \{x \in A: f(x) \neq 0\}$. Note that $f(x) \neq f(a)$ and thus there is a $b \in B$ for which $b(x) \neq b(a) = 0$. Therefore condition 2 of Stone's theorem is met. Let $\varepsilon > 0$ be given. By Stone's theorem there are a positive integer k, elements $b_1, \ldots, b_k$ of B, and a linear-ring map R such that $\|f(x) - R[B(x)]\| < \varepsilon/2$ for $x \in A$, where $B(x) = [(b_1 s)(x), \ldots, (b_k s)(x)]$. Since the b_j are continuous and A is

---

[38] Stone's theorem is an improved version of the well-known Stone–Weierstrass theorem, which, while very important, provides only sufficient conditions for approximation. The theorem, in the form above, was used earlier in Sandberg and Xu (1996). Theorem 5 of Stone (1962) gives a necessary and sufficient condition under which a continuous real-valued function f defined on a compact topological space X belongs to the family of functions generated by the linear-ring operations and uniform passage to the limit, starting from a given set χ_0 of continuous real-valued functions on X. The condition is that f must satisfy every linear relation of the form $g(x) = 0$ or $g(x) = g(y)$ that is satisfied by all functions in χ_0.

compact, $B(A)$ is compact. Choose $N \in D_k$ so that $|N(v) - R(v)| < \varepsilon/2$ for $v \in B(A)$, and observe that $|f(x) - N[B(x)]| < \varepsilon$ for $x \in A$.

Now suppose that statement 1 of the lemma is met but statement 2 is not. Then there are $a_1, a_2 \in A$ such that $f(a_1) \neq f(a_2)$ and $b(a_1) = b(a_2)$ for all b. Let $\delta > 0$ be such that $|f(a_1) - f(a_2)| > \delta$. Choose k, elements $b_1, \ldots, b_k$ of B, and an $N \in D_k$ such that $|f(a_1) - N[B(a_1)]| < \delta/2$ and $|f(a_2) - N[B(a_2)]| < \delta/2$. This [using $b(a_1) = b(a_2)$ for all b] gives the contradiction that

$$|f(a_1) - f(a_2)| \leq |f(a_1) - N[B(a_1)]| + |N[B(a_2)] - f(a_2)| < \delta$$

which completes the proof of the lemma.

Continuing with the proof of Theorem 4, suppose that (i) holds and choose any $n \in \mathbf{Z}_+$. Then, using the causality of G and the elements of H, the continuity of N, the continuity of the $q(h,n,\cdot)$, and the continuity of $(W_{n,\alpha})$, since the real-valued function F_n can be uniformly approximated arbitrarily well over S_n by continuous functions, F_n is continuous [see, e.g., Sutherlund (1975, pp. 120–121)].

Since G and the elements of H are time invariant, these maps take the zero element of S into the zero element of the set of $\mathbb{R}$-valued functions on $\mathbf{Z}_+$. Thus for each n, $F_n(\theta_n) = 0$ and $q(h,n,\theta_n) = 0$ for $h \in H$, where $\theta_n = (0, \ldots, 0)$. In addition, for each $\varepsilon > 0$ there are α, k, elements $h_1, \ldots, h_k$ of H, and an $N \in D_k$ such that

$$|F_n v - N(Qv)| < \varepsilon$$

for all $v \in S_n$, where $Qv = [q(h'_1,n,v), \ldots, q(h'_k,n,v)]$ and the h'_j are elements of H that satisfy $(h'_j s)(n) = (h_j W_{n,\alpha} s)(n)$ for each j. Thus, by Lemma 1, A.1 holds.

Assume now that (ii) is met, and let $\varepsilon > 0$ be given. Choose $\alpha \in \mathbf{Z}_+$ so that

$$|(Gs)(n) - (GW_{n,\alpha}s)(n)| < \varepsilon/2, \qquad n \in \mathbf{Z}_+ \qquad (2.15)$$

for $s \in S$.

By the continuity of F_α and the compactness of S_α, and by our lemma, there are k, elements $h_1, \ldots, h_k$ of H, and an $N \in D_k$ such that

$$|F_\alpha v - N(Pv)| < \varepsilon/2 \qquad (2.16)$$

for all $v \in S_\alpha$, where $Pv = [q(h_1, \alpha, v), \ldots, q(h_k, \alpha, v)]$. Now let $s \in S$ and $n \in Z_+$ be given. Suppose first that $n \geq \alpha$. By the time invariance of G and the h_j,

$$(GW_{n,\alpha}s)(n) - N[(MW_{n,\alpha}s)(n)] = F_\alpha v - N(Pv)$$

with M as described in the theorem and $v(\beta) = s(n - \alpha + \beta)$ for $\beta = 0, 1, \ldots, \alpha$. By (2.16)

$$|(GW_{n,\alpha}s)(n) - N[(MW_{n,\alpha}s)(n)]| < \varepsilon/2 \tag{2.17}$$

For $n < \alpha$,

$$(GW_{n,\alpha}s)(n) - N[(MW_{n,\alpha}s)(n)] = (Gs)(n) - N[(Ms)(n)] = F_\alpha v - N(Pv)$$

where now $v(\beta) = 0$ for $\beta < (\alpha - n)$ and $v(\beta) = s(\beta + n - \alpha)$ for $\beta = (\alpha - n), \ldots, \alpha$, showing that (2.17) holds even if $n < \alpha$. This together with (2.15) gives

$$|(Gs)(n) - N[(MW_{n,\alpha}s)(n)]| < \varepsilon$$

which completes the proof of the theorem.

Our next result is a corollary of Theorem 4. It focuses attention on the conditions required of the preprocessing stage represented by the h_j in Fig. 2.3 so that universal approximation is achieved. We will find that a separation condition is the key condition.

Theorem 6.   Let G be the set of all time-invariant causal approximately finite-memory maps G from S to the set of real-valued functions on Z_+ such that each associated functional F_n is continuous. Then (i) of Theorem 4 holds for each $G \in G$ if and only if for every n the $q(h, n, \cdot)$ separate the points of S_n [i.e., if and only if $q(h, n, u_1) \neq q(h, n, u_2)$ for some $h \in H$ whenever $u_1, u_2 \in S_n$ and $u_1 \neq u_2$].

Proof. The "if part" follows from Theorem 4.

Suppose now that (i) of Theorem 1 is satisfied for each $G \in G$, choose any $n \in Z_+$, and let u_1, u_2 be elements of S_n with $u_1 \neq u_2$. Observe that for any $j = 1, \ldots, d$ and any $\beta \in Z_+$ the map $(T_\beta \cdot)_j$ defined on S by $(T_\beta s)_j(\alpha) = [(T_\beta s)(\alpha)]_j$, in which $[(T_\beta s)(\alpha)]_j$ is the jth component of $(T_\beta s)(\alpha)$, is an element of G. The corresponding F_n is defined by $F_n u = [(T_\beta s)(n)]_j$ with s any element of S whose restriction to S_n is u. With j and β such that $(n - \beta) \geq 0$ and $[u_1(n - \beta)]_j \neq [u_2(n - \beta)]_j$, we have $F_n u_1 \neq F_n u_2$, and therefore, by Theorem 4, there is an $h \in H$ for which $q(h,n,u_1) \neq q(h,n,u_2)$. This shows that the $q(h,n, \cdot)$ separate the points of S_n.

2.4.3 Comments

For a given G the condition of Theorem 4 that A.1 be met can be much less restrictive than the separation condition of Theorem 6. For instance, suppose that G is the zero map [i.e., suppose that $(Gs)(n) = 0$ for all s and n]. Then each E_n is empty and A.1 imposes no restrictions on H, as one would expect.

Using the closure of H under the memory-limiting operation, a check of the proof of Theorem 1 shows that statement (i) is equivalent to: For each $\varepsilon > 0$, there are a positive integer k, elements $h_1, \ldots, h_k$ of H, and an $N \in D_k$ such that

$$\|(Gs)(n) - N[(Ms)(n)]\| < \varepsilon \qquad n \in \mathbf{Z}_+$$

for all $s \in S$, where $(Ms)(n) = [(h_1 s)(n), \ldots, (h_k s)(n)]$.

The conditions of statement (ii) of Theorem 4 can be expressed in other ways. For example, using the causality of G, it is not difficult to check that the condition that each F_n is continuous is equivalent to the condition that each functional $G(\cdot)(n)$ is continuous with respect to the metric on S given by $\rho(x, y) = \sup\{\|x(j) - y(j)\|: j \in \mathbf{Z}_+\}$. As another example, the statement that each F_n is continuous contains redundancy in the sense that, by the time invariance of G, if n_1 and n_2 are elements of $\mathbf{Z}_+$ such that $n_2 > n_1$, then the continuity of F_{n2} implies the continuity of F_{n1}. Similarly, and concerning A.1 (and this time by the time invariance of G and the elements of H), if the following holds for $n = n_2$ and $n_2 > n_1$, then it holds for $n = n_1$.

For each $(u_1, u_2) \in E_n$ there is an $h \in H$ such that $q(h,n,u_1) \neq q(h,n,u_2,)$, in which

$$E_n = \{(u_1, u_2) \in S_n \times S_n : F_n u_1 \neq F_n u_2\}$$

In applications it is often possible to choose H so that its elements are linear and A.1 is met. However, for a given degree of approximation (i.e., for a given ε) a much lower overall degree of complexity of the approximation structure can sometimes result if H is allowed to contain relatively simple elements that are not linear. As an example of how H can be chosen, let H_{00} stand for the subset of H_0 (of Section 2.1) containing only the members for which the functions ϕ are strictly monotone increasing. Given n as well as u_1 and u_2 in S_n such that $u_1 \neq u_2$, choose $h \in H_{00}$, so that it is given by (2.14) with $a(n - j) = (u_1 - u_2)(j)^{tr}$ for $j = 1, \ldots, n$. Then

$$\sum_{j=0}^{n} (u_1 - u_2)(j)a(n - j) \neq 0$$

and so, by the strict monotonicity of the ϕ's, $q(n,h,u_1) \neq q(n,h,u_2)$. This shows that A.1 is always met for $H = H_{00}$.

Now assume that $H = H_{00}$ with the ϕ's linear and assume also that the D_k are linear-ring maps. Let Q denote the approximating map of Theorem 4 given by $(Qs)(n) = N[(MW_{n,\alpha}s)(n)]$, and let $h'_1, \ldots, h'_k$ be elements of H such that $(h'_j s)(n) = (h_j W_{n,\alpha}s)(n)$. Consider the $d = 1$ case. By writing products of sums as iterated sums, it is not difficult to see that $(Q_s)(n)$ has the form

$$\sum_{j=1}^{p} \sum_{n_j=0}^{n} \ldots \sum_{n_1=0}^{n} k_j(n - n_1, \ldots, n - n_j)s(n_1) \ldots s(n_j) \qquad (2.18)$$

in which p is a positive integer, and each $k_j(n - n_1, \ldots, n - n_j)$ is a finite linear combination of products of the form $a'_{i_j(1)}(n - n_1)a'_{i_j(2)}(n - n_2) \ldots a'_{i_j(j)}(n - n_j)$, where the indices $i_j(1), \ldots, i_j(j)$ are drawn from $\{1, \ldots, k\}$ and each a'_i is the kernel (i.e., discrete impulse response) associated with h'_j. Note that each k_j vanishes whenever one or more of its arguments exceed α.

Expression (2.18) is a doubly finite Volterra series approximant for G in the sense that p is finite and each k_j vanishes if at least one of its arguments is sufficiently large in the sense indicated (which, of course, implies that the approximants have finite memory). Let $\mathcal{M}$ stand for the set of all causal time-invariant maps from S to the real-valued functions defined on $\mathbf{Z}_+$, and let $\mathcal{V}$ denote the family of all members of $\mathcal{M}$ that have a representation of the form (2.18) for some p and some $k_1, \ldots, k_p$ such that for each j we have $k_j(n_1, \ldots, n_j) = 0$ when one or more of the $n_1, \ldots, n_j$ exceed some number η. We have just seen that sufficient conditions for a member of $\mathcal{M}$ to be uniformly approximated arbitrarily well by an element of $\mathcal{V}$ are that the member possesses approximately finite memory and that its corresponding F_n functionals are continuous. In Appendix I we show that these conditions are in fact necessary.[39,40]

The methods we have described can be used to obtain corresponding results for other types of input–output maps. A particularly important case is the one in which inputs and outputs are real-valued functions of a finite number of integer-valued variables. This case is of interest in connection with, for example, image processing. We consider this case in the next section, where a large class of myopic maps are the focus of attention. As an application, we give criteria in this multivariable setting for the existence of arbitrarily good doubly finite Volterra series approximations.

2.4.4 Approximation and Discrete-Space Myopic Systems

We first introduce symbols U, K, and L, which play roles analogous to S, G, and H, respectively.

Let m be a positive integer, and let U denote the set of all maps u from $\mathbf{Z}^m$ to C ($\mathbf{Z}^m$ is the set of m vectors with integer components, and C is defined in Section 2.1). Here U is our set of inputs.

For each $\beta \in \mathbf{Z}^m$ and each $r \in \mathbf{Z}_+$, let maps $W_\beta^r: U \to U$ and $T^\beta: U \to U$ be defined by

$$(W_\beta^r u)(\gamma) = \begin{cases} u(\gamma) & \max_j |\gamma_j - \beta_j| \le r \\ 0 & \text{otherwise} \end{cases}$$

(where γ_j and β_j denote the jth components of γ and β, respectively) and

$$(T^\beta u)(\gamma) = u(\gamma - \beta)$$

---

[39] Also, we have already mentioned (in a footnote in Section 2.4.1) that the meaning of "approximately finite memory" is different in Sandberg (1991), where α is required to be positive instead of nonnegative. It is natural to ask how Theorems 4, 6, and 9 (Theorem 9 is in Appendix I) might change if we had used the definition in Sandberg (1991). A quick check of the proofs shows that, except for replacing "$\alpha \in \mathbf{Z}_+$" with "positive $\alpha \in \mathbf{Z}_+$" in Theorem 1, they do not change at all.

[40] For related material concerning the quality of approximation that can be achieved using approximants of a given order, see Sandberg (2000) as well as Sandberg (1999a).

We say that a map M from U into the set of real-valued functions on $\mathbf{Z}^m$ is *shift invariant* if for each $\beta \in \mathbf{Z}^m$ we have $(MT^\beta u)(\gamma) = (Mu)(\gamma - \beta)$ for all γ and u.

Throughout Section 2.4.4, K denotes a shift-invariant map from U to the set of real-valued functions defined on $\mathbf{Z}^m$. We assume that $K\theta = \theta$ (here θ stands for the zero element of U as well as the zero element of the set of real-valued functions on $\mathbf{Z}^m$) and that K is *myopic* in the sense that given $\varepsilon > 0$ there is an $r \in \mathbf{Z}_+$ such that

$$|(Ku)(\gamma) - (KW_\gamma^r u)(\gamma)| < \varepsilon \qquad \gamma \in \mathbf{Z}^m$$

for $u \in U$.

For each $n \in \mathbf{Z}_+$, let c_n in this section stand for $\{-n, \ldots, 0, \ldots, n\}^m$, and let U_n denote the restriction of U to c_n. We view each U_n as a metric space with metric ρ_n defined by $\rho_n(x,y) = \max\{\|x(j) - y(j)\|: j \in c_n\}$ (recall that $\|\cdot\|$ is any norm on $\mathbb{R}^d$). We shall use the fact that each U_n is compact. We view U as a metric space with metric ρ given by $\rho(x,y) = \sup\{\|x(j) - y(j)\|: j \in \mathbf{Z}^m\}$.

Let L be a family of shift-invariant maps l from U to the set of $\mathbb{R}$-valued functions defined on $\mathbf{Z}^m$ such that $l\theta = \theta$ for each l, and for each l and each n in $\mathbf{Z}_+$ define the functional $q(l,n,\cdot)$ on U_n by $q(l,n,v) = (lu)(0)$, where u is the element of U given by $u(\gamma) = v(\gamma)$ for $\gamma \in c_n$ and $u(\gamma) = 0$ otherwise. We assume that $q(l,n,\cdot)$ is continuous for each l and each n. We also assume that L is closed under the windowing operation, in the sense that g defined on U by $(gu)(\gamma) = (lW_\gamma^r u)(\gamma)$ belongs to L whenever $l \in L$ and $r \in \mathbf{Z}_+$.[41]

As an example, we can take L to be the set L_0 of all maps l for which

$$(lu)(\gamma) = \phi\left[\sum_{\beta \in \mathbf{Z}^m} u(\beta)a(\gamma - \beta)\right] \qquad \gamma \in \mathbf{Z}^m \qquad (2.19)$$

where ϕ, which depends on l, is a continuous map from $\mathbb{R}$ into $\mathbb{R}$ with $\phi(0) = 0$, and a, which also depends on l, is real $d \times 1$ matrix valued with $a(\beta)$ the zero $d \times 1$ matrix for $\max_j\{\beta_j\}$ sufficiently large. As another example, note that L can be taken to be any subset of L_0 that is closed under the windowing operation.

---

[41] In this connection, g defined above is shift invariant for $l \in L$ and $r \in \mathbf{Z}_+$.

For each $n \in \mathbf{Z}_+$, let J_n denote the functional defined on U_n by $J_n v = (Ku)(0)$, where u is the element of U given by $u(\gamma) = v(\gamma)$ for $\gamma \in c_n$ and $u(\gamma) = 0$ otherwise. We shall use A.2 to denote the following condition:

For each $n \in \mathbf{Z}_+$ and each $(u_1, u_2) \in E_n$ there is an $l \in L$ such that $q(l,n,u_1) \neq q(l,n,u_2)$,

in which

$$E_n = \{(u_1, u_2) \in U_n \times U_n : J_n u_1 \neq J_n u_2\}$$

Our result analogous to Theorem 4 is:

Theorem 7. The following two statements are equivalent.

(i) For each $\varepsilon > 0$, there are an $r \in \mathbf{Z}_+$, a positive integer k, elements $l_1, \ldots, l_k$ of L, and an $N \in D_k$ such that

$$|(Ku)(\gamma) - N[(MW_\gamma^r u)(\gamma)]| < \varepsilon \qquad \gamma \in \mathbf{Z}^m$$

for all $u \in U$, where $(Mu)(\gamma) = [(l_1 u)(\gamma), \ldots, (l_k u)(\gamma)]$.

(ii) Each J_n is continuous and A.2 is met.

The proof, which parallels the proof of Theorem 4, is given in Appendix J.[42]

Our next result is a corollary of Theorem 4 that parallels Theorem 3. In the setting of this section, it focuses attention on the conditions required of the preprocessing stage in Fig. 2.3 (with the h_j replaced with l_j, and with x and n_0 replaced with s and k, respectively) so that universal approximation is achieved. Again we find that a separation condition is the key condition.

Theorem 8. Let $\mathcal{K}$ be the set of all shift-invariant myopic maps K from U to the set of real-valued functions on $\mathbf{Z}^m$ such that each associated functional J_n is continuous and $K\theta = \theta$. Then (i) of Theorem

---

[42] And using the closure of H under the windowing operation, a check of the proof of Theorem 7 shows that its statement (i) is equivalent to: For each $\varepsilon > 0$, there are a positive integer k, elements $l_1, \ldots, l_k$ of L, and an $N \in D_k$ such that $|(Ku)(\gamma) - N[(Mu)(\gamma)]| < \varepsilon$ $(\gamma \in \mathbf{Z}^m)$ for all $u \in U$, where $(Mu)(\gamma) = [(l_1 u)(\gamma), \ldots, (l_k u)(\gamma)]$.

4 holds for each $K \in \mathcal{K}$ if and only if for every n the $q(l,n,\cdot)$ separate the points of U_n [i.e., if and only if $q(l,n,u_1) \neq q(l,n,u_2)$ for some $l \in L$ whenever $u_1, u_2 \in U_n$ and $u_1 \neq u_2$].

The proof of this theorem is essentially the same as the proof of Theorem 6, with the map $(T^\beta \cdot)_j$ defined on U by $(T^\beta u)_j(\gamma) = [(T^\beta u)(\gamma)]_j$ playing the role of $(T_\beta \cdot)_j$ in the proof of Theorem 6. The details are easily filled in and so are omitted.

In Appendix K we show that for $d = 1$ every element of $\mathcal{K}$ can be approximated uniformly arbitrarily well by a doubly finite Volterra series of the form

$$\sum_{j=1}^{p} \sum_{\beta_j \in \mathbf{Z}^m} \cdots \sum_{\beta_1 \in \mathbf{Z}^m} k_j(\gamma - \beta_1, \ldots, \gamma - \beta_j) u(\beta_1) \ldots u(\beta_j) \qquad (2.20)$$

It is also true that if a K can be so approximated, then it is myopic and each associated functional J_n is continuous. Similar results hold also for general d. In the general case the k_j are row-vector valued and $u(\beta_1) \ldots u(\beta_j)$ stands for a Kronecker product.[43]

2.5 CONCLUDING COMMENTS

For a variety of reasons there is a pressing need to understand the behavior of nonlinear systems and to establish in different contexts an analytical basis for their design. As is well known, the analysis of a nonlinear system using current mathematical methods can be a formidable task. One common alternative approach is to employ simulation techniques. While such techniques are often relatively easy to employ and can be useful, they possess serious well-known limitations. In this chapter we have described an ongoing study based on the use of concepts drawn from the areas of real analysis and functional analysis. These concepts are typically more difficult to master, but they are powerful and help to provide useful insights.

We have considered classification problems and the related problem of approximating dynamic nonlinear input–output maps. Attention is focused on the properties of nonlinear structures that

---

[43] Although not mentioned earlier, such similar results for general d hold also in the setting of Section 2.4.3.

have the form of a dynamic preprocessing stage followed by a memoryless nonlinear section. We were able to show that an important family of classification problems can be solved using certain simple structures involving linear functionals and memoryless nonlinear elements. While natural questions arise concerning specific practical problems and detailed implementations, such questions are not addressed in this chapter. We make no apologies for not having considered these questions. We have been concerned mainly with questions concerning what is possible and what is not. Answers to such questions are often of considerable value.

We have considered the problem of approximating discrete-space multidimensional shift-invariant input–output maps with inputs drawn from a certain large set. Our main result in this setting is a theorem showing that if the maps to be approximated satisfy a certain "myopicity" condition, which is very often met, then they can be uniformly approximated arbitrarily well using a structure consisting of a linear preprocessing stage followed by a memoryless nonlinear network. Noncausal as well as causal maps are considered. As mentioned earlier, approximations for noncausal maps for which inputs and outputs are functions of more than one variable are of current interest in connection with, for instance, image processing. We have also considered other aspects of the problem of approximating nonlinear input–output maps in which, for example, certain separation conditions play a central role. In the course of a study such as ours, interesting unexpected side issues sometimes arise. One such issue is discussed in Appendix G where, in connection with our study of myopic maps, we were led to consider the representation of linear discrete-time maps and have focused attention on—and corrected—a long-standing oversight concerning the cornerstone of digital signal processing.

2.6 APPENDICES

Appendix A: Completeness of X_w of Section 2.3.1

To see that X_w is complete, let $\{x_j\}$ be a Cauchy sequence in X_w[44] and notice that for each positive ε there is a positive N such that

---

[44] Recall that $\{x_j\}$ is Cauchy in X_w if each x_j belongs to X_w and for each $\varepsilon > 0$ there is a positive integer N for which $\|x_p - x_q\|_w < \varepsilon$ for $p, q > N$.

$$\|wx_p - wx_q\|_0 < \varepsilon \qquad p,q > N$$

where $\|\cdot\|_0$ is the norm in $X(\mathbf{Z}^m,\mathbb{R}^n)$ given by $\|x\|_0 = \sup_{\alpha \in \mathbf{Z}^m}\|x(\alpha)\|$. By the completeness of $X(\mathbf{Z}^m,\mathbb{R}^n)$, there is a $y \in X(\mathbf{Z}^m,\mathbb{R}^n)$ such that $\{wx_j\}$ converges to y in $X(\mathbf{Z}^m,\mathbb{R}^n)$. Thus, for an arbitrary positive ε there exists a positive $\tilde{N}$ such that

$$\|y - wx_j\|_0 < \varepsilon \qquad j > \tilde{N}$$

Since w never vanishes, the function y/w can be seen to be the limit of $\{x_j\}$ in X_w.

Appendix B: Absolute Summability

Here, for the reader's convenience, we describe certain facts that we use concerning absolutely summable families of functions. The material is essentially standard.[45] As in Section 2.3.1, let $D = \mathbf{Z}^m$ or $\mathbf{Z}_-^m$. Let f be a map from D to E, where E is either $\mathbb{R}$ or $\mathbb{R}^n$. Let $\|\cdot\|_*$ denote the absolute value operator if $E = \mathbb{R}$, and let $\|\cdot\|_* = \|\cdot\|$ if $E = \mathbb{R}^n$.

We say that f belongs to $S_1(D)$ if there is a positive constant c such that

$$\sum_{\beta \in J} \|f(\beta)\|_* \le c$$

for all finite subsets J of D. Such an f is said to be *absolutely summable* on D, and another way to denote this is to write

$$\sum_{\beta \in D} \|f(\beta)\|_* < \infty$$

If $f \in S_1(D)$, then $\sum_{|\beta| \le k} f(\beta)$ converges in E to a limit l as $k \to \infty$, and we denote this l by $\sum_{\beta \in D} f(\beta)$.

Write $\beta = (\beta_1, \ldots, \beta_m)$, and for each j let $\sum_{\beta_j}$ denote a sum in which β_j ranges over $\mathbf{Z}$ if $D = \mathbf{Z}^m$, or over $\mathbf{Z}_-$ if $D = \mathbf{Z}_-^m$. If $f \in S_1(D)$, the

---

[45] It can be viewed as a direct modification of material in Dieudonné (1969, pp. 97–100) or as a special case of the general theory of integration with respect to a measure.

iterated sum $\Sigma_{\beta_m} \ldots \Sigma_{\beta_1} f(\beta)$ is well defined in the sense that each sum converges absolutely, and we have

$$\sum_{\beta \in D} f(\beta) = \sum_{\beta_m} \ldots \sum_{\beta_1} f(\beta) \tag{2.21}$$

It is also true that (2.21) holds if $\beta_m, \ldots, \beta_1$ is replaced with any permutation $\beta_{q(m)}, \ldots, \beta_{q(1)}$ of $\beta_m, \ldots, \beta_1$.

Appendix C: On $G(w)$ and $G_-(w)$ of Section 2.3.1

Let $D = \mathbf{Z}^m$ or $\mathbf{Z}_-^m$. Denote by A the set of all functions g from D to $\mathbb{R}^n$ satisfying (2.6), and recall that these functions are absolutely summable. Let B be any dense subset of A, in the sense that for each $\varepsilon > 0$ and each $g \in A$ there is an $f \in B$ such that

$$\sum_{\beta \in D} \|g(\beta) - f(\beta)\| < \varepsilon$$

It is clear that any dense subset of $l_1(D)$[46] whose elements satisfy (2.6) is a dense subset of A.[47] We now show that $G(w)$ and $G_-(w)$ can be taken to be B when $D = \mathbf{Z}^m$ and $D = \mathbf{Z}_-^m$, respectively.

Let $x \in X(\mathbf{Z}^m, \mathbb{R}^n)$ be such that the restriction of x to D is nonzero. We are going to show that there is an $f \in B$ such that the sum $\Sigma_{\beta \in D} < f(\beta), x(\beta) >$ is not zero. Choose any $r \in l_1(D)$ that never vanishes, and let g be the element of A given by

---

[46] For $D = \mathbf{Z}^m$ or $\mathbf{Z}_-^m$, $l_1(D)$ denotes the normed space of all $r: D \to \mathbb{R}^n$ such that r is absolutely summable on D, with the norm $\|\cdot\|_1$ in $l_1(D)$ given by $\|r\|_1 = \Sigma_{\beta \in D} \|r(\beta)\|$. In other words, $l_1(D)$ denotes the set $S_1(D)$ of Appendix B with $E = \mathbb{R}^n$ and with the norm indicated here.

[47] For example, suppose that $D = \mathbf{Z}_-^m$, that $n = m = 1$, and that σ and λ are positive constants such that $\lambda > \sigma$ and $w(\beta)^{-1} = O(e^{-\sigma \beta})$. Then B can be taken to be the set of all finite linear combinations of the functions $k_1, k_2, \ldots$ where $k_j(\beta) = (\beta)^{(j-1)} e^{\lambda \beta}$ for each j. [A proof of the denseness of this B in $l_1(\mathbf{Z}_-)$ can be obtained by modifying material in Stone (1962, pp. 75–76) concerning a related continuous-time case.] As another example, assume that $D = \mathbf{Z}^m$, that $n = m = 1$, and that σ and λ are positive constants such that $\lambda > \sigma$ and $w(\beta)^{-1} = O(e^{\sigma \beta_2})$. In this case B can be taken to be the set of all finite linear combinations of the functions $k_1, k_2, \ldots$ where $k_j(\beta) = (\beta)^{(j-1)} e^{-\lambda \beta^2}$ for each j [see Stone (1962, p. 80) for a corresponding continuous-time denseness result]. As a third example, for $D = \mathbf{Z}^m$ we can take B to be the family of all finite linear combinations of the elements of the set $\{e_\alpha^j\}$ described in an earlier footnote.

$$g(\beta) = \|r(\beta)\|w^2(\beta)x(\beta) \qquad \beta \in D$$

Using

$$\sum_{\beta \in D} \langle g(\beta), x(\beta) \rangle = \sum_{\beta \in D} \|r(\beta)\| \cdot \|w(\beta)x(\beta)\|^2 > 0$$

select $f \in B$ such that

$$2 \sup_{\beta \in \mathbf{Z}^m} \|x(\beta)\| \cdot \sum_{\beta \in D} \|f(\beta) - g(\beta)\| < \sum_{\beta \in D} \langle g(\beta), x(\beta) \rangle$$

Thus,

$$\left| \sum_{\beta \in D} \langle f(\beta), x(\beta) \rangle \right| \geq \sum_{\beta \in D} \langle g(\beta), x(\beta) \rangle - \sum_{\beta \in D} \|f(\beta) - g(\beta)\| \cdot \|x(\beta)\|$$

showing that

$$\left| \sum_{\beta \in D} \langle f(\beta), x(\beta) \rangle \right| > \frac{1}{2} \sum_{\beta \in D} \langle g(\beta), x(\beta) \rangle > 0$$

Appendix D: Proof of Lemma 2

Let $cl(U)$ denote the closure of U in X_w. We shall use the proposition that $cl(U) \subset X(\mathbf{Z}^m, \mathbb{R}^n)$, which follows from the uniform boundedness of U and the fact that convergence in X_w implies uniform convergence on bounded subsets of $\mathbf{Z}^m$.

Since $cl(U)$ is closed, F has a continuous extension F_e to $cl(U)$ [Rudin (1976, p. 99, Problem 13)]. Let A be the set of all real-valued functions a on $cl(U)$ given by

$$a(u) = \rho + c \sum_{\beta \in \mathbf{Z}^m} \langle g(\beta), u(\beta) \rangle \tag{2.22}$$

with $g \in \mathcal{G}(w)$ and ρ and c real numbers. By the Schwarz inequality in $\mathbb{R}^n$, and the observation that $\langle g(\beta), u(\beta) \rangle = \langle w(\beta)^{-1}g(\beta), w(\beta)u(\beta) \rangle$, each a is continuous. Let u_1 and u_2 be elements of $cl(U)$, and let r_1

and r_2 be real numbers such that $r_1 = r_2$ if $u_1 = u_2$. If $u_1 = u_2$, choose $c = 0$. If u_1 and u_2 are distinct, pick $g \in G(w)$ and c such that

$$c \sum_{\beta \in \mathbf{Z}^m} \langle g(\beta), u_1(\beta) - u_2(\beta) \rangle = r_1 - r_2$$

Observe that with $\rho = r_1 - c\Sigma_{\beta \in \mathbf{Z}^m}\langle g(\beta), u_1(\beta)\rangle$, we have $a(u_1) = r_1$ and $a(u_2) = r_2$, where a is given by (2.22). We will use the following lemma, which is a direct consequence of Stone (1962, p. 35, Theorem 1).

Lemma. Let X be the family of continuous real-valued functions on $\mathrm{cl}(U)$, let X_0 be an arbitrary subfamily of X, and let L_0 be the family of all functions generated from X_0 by the lattice operations on real-valued functions and uniform passage to the limit. Suppose that $x \in X$, and that for any u_1 and u_2 in $\mathrm{cl}(U)$ and any real numbers r_1 and r_2 such that $r_1 = r_2$ if $u_1 = u_2$ there exists an $a \in X_0$ such that $a(u_1) = r_1$ and $a(u_2) = r_2$. Then $x \in L_0$.

Let $\varepsilon > 0$ be given. By the lemma, with $X_0 = A$ and $x = F_e$, there are n_0, elements $a_1, \ldots, a_{n_0}$ of A, and a lattice map[48] $M: \mathbb{R}^{n_0} \to \mathbb{R}$ such that

$$|F_e(u) - M[a_1(u), \ldots, a_{n_0}(u)]| < \varepsilon/2$$

for $u \in \mathrm{cl}(U)$. For $j = 1, \ldots, n_0$ let ρ_j, c_j, and g_j be the ρ, c, and g associated with a_j, respectively. For each j define y_j on $\mathrm{cl}(U)$ by

$$y_j(u) = \sum_{\beta \in \mathbf{Z}^m} \langle g_j(\beta), u(\beta) \rangle$$

and notice that each $K_j := y_j(\mathrm{cl}(U))$ is a compact set in $\mathbb{R}$. Let $N \in D_{n_0}$ satisfy

$$|M(c_1 v_1 + \rho_1, \ldots, c_{n_0} v_{n_0} + \rho_{n_0}) - N(v_1, \ldots, v_{n_0})| < \varepsilon/2$$

for $v := (v_1, \ldots, v_{n_0}) \in (K_1 \times \ldots \times K_{n_0})$. Thus, for any $u \in U$

---

[48] The *lattice operations* $a \vee b$ and $a \wedge b$ on pairs of real numbers a and b are defined by $a \vee b = \max(a,b)$ and $a \wedge b = \min(a,b)$. We say that a map $M: \mathbb{R}^{n_0} \to \mathbb{R}$ is a *lattice map* if Mv is generated from the components $v_1, \ldots, v_{n_0}$ of v by a finite number of lattice operations that do not depend on v.

$$|F(u) - N[y_1(u), \ldots, y_{n_0}(u)]|$$
$$\leq |F(u) - M[a_1(u), \ldots, a_{n_0}(u)]|$$
$$+ |M[a_1(u), \ldots, a_{n_0}(u)] - N[y_1(u), \ldots, y_{n_0}(u)]| < \varepsilon/2 + \varepsilon/2$$

as required.

Another way to prove Lemma 2 is to make use of the Stone–Weierstrass theorem instead of the lemma described above. The details are almost the same. The relationship in Stone (1962) between (Stone 1962, p. 35, Theorem 1) and the Stone–Weierstrass theorem is that Stone (1962, p. 35, Theorem 1) is used in the proof of a version of the Stone–Weierstrass theorem. Of course the proof given above shows that for our purposes it is not necessary to use the Stone–Weierstrass theorem, and it focuses attention on the important concept of a lattice map.

Appendix E: Proof of Lemma 3

Let $\text{cl}(QU)$ denote the closure of QU in X_w. Since $QU \subseteq U$, QU is relatively compact in X_w and (by the comments at the beginning of Appendix D) $\text{cl}(QU) \subseteq \text{cl}(U) \subset X(\mathbf{Z}^m, \mathbb{R}^n)$. In particular, the map F_e described in Appendix D is continuous on $\text{cl}(QU)$ with respect to $\|\cdot\|_w$. Here let A be the set of all real-valued functions a on $\text{cl}(QU)$ defined by

$$a(u) = \rho + c \sum_{\beta \in \mathbf{Z}^m} \langle g(\beta), u(\beta) \rangle$$

with $g \in G_-(w)$ and ρ and c real numbers. Here too each a is continuous. Let u_1 and u_2 be distinct elements of $\text{cl}(QU)$, and let r_1 and r_2 be arbitrary real numbers. Then $(u_1 - u_2) \in X(\mathbf{Z}^m, \mathbb{R}^n)$ is nonzero and, by the definition of Q, the restriction $(u_1 - u_2)|_{\mathbf{Z}_-^m}$ of $(u_1 - u_2)$ to $\mathbf{Z}_-^m$ is nonzero. Hence there is an $a \in A$ such that $a(u_1) = r_1$ and $a(u_2) = r_2$, and obviously the same conclusion holds even if u_1 and u_2 are not distinct provided that $r_1 = r_2$. This establishes the applicability of the lemma in Appendix D, with $X_0 = A$ and $x = F_e$ restricted to $\text{cl}(QU)$. The remainder of the proof is essentially the same as that of Lemma 1. Here the y_j are defined on $\text{cl}(QU)$, and

$$y_j(u) = \sum_{\beta \in \mathbf{Z}_-^m} \langle g_j(\beta), u(\beta) \rangle$$

Appendix F: Myopic Maps and Causality

Let R be any map from S to $X(\mathbf{Z}^m, \mathbb{R}^n)$ such that $(Rs)(\alpha) = s[\gamma(\alpha)]$, where $\gamma\colon \mathbf{Z}^m \to \mathbf{Z}_-^m$ satisfies $\gamma(\alpha) = \alpha$ for $\alpha \in \mathbf{Z}_-^m$. For example, we can take R to be the Q of Section 2.3.1 or the Q defined by (2.10).

Proposition 1. Let S be a uniformly bounded subset of $X(\mathbf{Z}^m, \mathbb{R}^n)$. If $RS \subseteq S$ and G is causal and myopic on S with respect to w, then G meets condition (2.11) with $\mu = w|\mathbf{Z}_-^m$.

Proof. Let $\varepsilon > 0$ be given, and let $\delta > 0$ be such that (2.4) is met. Let c be the uniform bound for S, and let $b_k = \{\alpha \in \mathbf{Z}^m : |\alpha| \le k\}$ for each $k \in \mathbb{N}$. Using $w(\alpha) \to 0$ as $|\alpha| \to \infty$, choose k such that $2c|w(\alpha)| < \delta$ for $\alpha \notin b_k$, and hence so that

$$\sup_{\alpha \notin b_k} \|w(\alpha)[(Rx)(\alpha) - (Ry)(\alpha)]\| < \delta$$

for any x and y in S. Let x and y satisfy

$$\max_{\alpha \in b_k}|w(\alpha)| \cdot \sup_{\alpha \in \mathbf{Z}_-^m}\|w(\alpha)[x(\alpha) - y(\alpha)]\| < \delta \min_{\beta \in d_k}|w(\beta)|$$

where $d_k = \{\gamma(\alpha)\colon \alpha \in b_k\}$. Then

$$\sup_{\alpha \in b_k}\|w(\alpha)[(Rx)(\alpha) - (Ry)(\alpha)]\|$$

$$\le \max_{\alpha \in b_k}|w(\alpha)| \cdot \max_{\beta \in d_k}\left|w(\beta)^{-1}\right| \cdot \sup_{\beta \in d_k}\|w(\beta)[x(\beta) - y(\beta)]\|$$

$$\le \max_{\alpha \in b_k}|w(\alpha)| \cdot \left(\min_{\beta \in d_k}|w(\beta)|\right)^{-1} \cdot \sup_{\alpha \in \mathbf{Z}_-^m}\|w(\alpha)[x(\alpha) - y(\alpha)]\|$$

$$< \delta$$

Since

$$\sup_{\alpha \in \mathbf{Z}^m}\|w(\alpha)[(Rx)(\alpha) - (Ry)(\alpha)]\| < \delta$$

and $RS \subseteq S$ and G is myopic, we have

$$|(GRx)(0) - (GRy)(0)| < \varepsilon$$

Using $(Rs)(\alpha) = s(\alpha)$ for all $s \in S$ and $\alpha \in \mathbf{Z}^m_-$, together with the causality of G, we have

$$|(Gx)(0) - (Gy)(0)| < \varepsilon$$

This completes the proof.

Proposition 2. Suppose that condition (2.11) is met. Then G is causal.[49]

Proof. Let $\alpha \in \mathbf{Z}^m$ be arbitrary, and suppose that x and y are two elements of S with $x(\beta) = y(\beta)$ for all $\beta_j \leq \alpha_j$. Clearly, $(T_{-\alpha}x)(\beta) = (T_{-\alpha}y)(\beta)$ for $\beta \in \mathbf{Z}^m_-$. By condition (2.11), $|(GT_{-\alpha}x)(0) - (GT_{-\alpha}y)(0)| < \varepsilon$ for all $\varepsilon > 0$, which gives $(GT_{-\alpha}x)(0) = (GT_{-\alpha}y)(0)$. Thus, using the shift invariance of G, we have $(Gx)(\alpha) = (Gy)(\alpha)$, which proves the proposition.

Finally, suppose that μ is as described in connection with condition (2.11), and let v be any extension of μ to $\mathbf{Z}^m$ that never vanishes and satisfies $\lim_{|\alpha| \to \infty} v(\alpha) = 0$. [For example, we can take $v(\alpha) = \mu(\beta)$ for all α, where $\beta_j = -|\alpha_j|$ for each j.] Since

$$\sup_{\alpha \in \mathbf{Z}^m_-} \|\mu(\alpha)[x(\alpha) - y(\alpha)]\| = \sup_{\alpha \in \mathbf{Z}^m_-} \|v(\alpha)[x(\alpha) - y(\alpha)]\|$$
$$\leq \sup_{\alpha \in \mathbf{Z}^m} \|v(\alpha)[x(\alpha) - y(\alpha)]\|$$

it follows directly that G is myopic with respect to v if condition (2.11) is met.

Appendix G: Representation Theorem for Linear Discrete-Space Systems

The material is this appendix, while related to the topic of myopic maps, is a self-contained section with its own notation and definitions.

The cornerstone of the theory of discrete-time single-input–single-output linear systems is the idea that every such system has an input–output map H that can be represented by an expression of the form

---

[49] This is observed in Boyd and Chua (1985) for $m = n = 1$.

$$(Hx)(n) = \sum_{p=-\infty}^{\infty} h(n, p)x(p) \tag{2.23}$$

in which x is the input and h is the system function associated with H in a certain familiar way. It is widely known that this, and a corresponding representation for time-invariant systems in which $h(n,p)$ is replaced with $h(n - p)$, are discussed in many books [see, e.g., Seibert (1997, pp. 267–269); Oppenheim et al. (1997, pp. 77–79); or Proakis and Manolakis (1992, pp. 46–71)]. Almost always it is emphasized that these representations hold *for all* linear input–output maps H. On the other hand, in Boyd and Chua (1985, p. 1159) attention is directed to material in Kantorovich and Akilov (1982, p. 58) that shows that certain time-invariant H's in fact do not have convolution representations.[50] This writer does not claim that these H's are necessarily of importance in applications, but he does feel that their existence shows that the analytical ideas in the books are flawed.[51]

One of the main purposes of this appendix is show that, under some mild conditions concerning the set of inputs and H, (2.23) becomes correct if an additional term is added to the right side. More specifically, we show that

$$(Hx)(n) = \sum_{p=-\infty}^{\infty} h(n, p)x(p) + \lim_{k \to \infty}(HE_k x)(n)$$

---

[50] The claim in Boyd and Chua (1985) that there exists a time-invariant causal H that has no convolution representation is correct, but it may not be clear that the argument given there actually shows this. Specifically, it may not be clear from what is said in Boyd and Chua (1985) that the pertinent map constructed there is causal and time invariant. However, it is not difficult to modify what is said so that it establishes the claim [see Kantorovich and Akilov (1982, p. 58) or the proof of the proposition in Appendix G]. It appears that as early as 1932 Banach was aware of the lack of existence of generalized-convolution representations for certain linear system maps [see Edwards (1995, pp. 158, 159)]. For related studies, see Borodziewicz et al. (1983) and Kishore and Pearson (1994).

[51] The oversight in the books is due to the lack of validity of the interchange of the order of performing a certain infinite sum and then applying $(H\cdot)(n)$. The infinite sum at issue clearly converges pointwise, but that is to enough to justify the interchange. A special case in which the interchange is justified is that in which the inputs are elements of $\ell_p (1 \le p \le \infty)$ and each $(H\cdot)(n)$ is a bounded linear functional on ℓ_p.

for each n, in which h has the same meaning as in (2.23), and $E_k x$ denotes the function given by $(E_k x)(p) = x(p)$ for $|p| > k$ and $(E_k x)(p) = 0$ otherwise. This holds whenever the input set is the set of bounded functions, the outputs are bounded, and H is continuous. In particular, we see that in this important setting, an H has a representation of the form given by (2.23) if and only if

$$\lim_{k \to \infty}(H E_k x)(n) = 0$$

for all x and n. Since this is typically a very reasonable condition for a system map H to satisfy, it is clear that the H's that cannot be represented using just (2.23) are rather special.

Our results concerning H are given in the following section, in which the setting is more general in that we address H's for which inputs and outputs depend on an arbitrary finite number of variables. We also consider H's for which inputs and outputs are defined on just the nonnegative integers. In that setting the situation with regard to the need for an additional term in the representation is different: No additional term is needed for causal maps H.

G.1 Linear System Representation Result

Preliminaries Let m be a positive integer, let $\mathbf{Z}$ be the set of all integers, and let $\mathbf{Z}_+$ denote the set of nonnegative integers. Let D stand for either $\mathbf{Z}^m$ or $\mathbf{Z}_+^m$. Let $\ell_\infty(D)$ denote the normed linear space of bounded $\mathbb{R}$-valued functions x defined on D, with the norm $\|\cdot\|$ given by $\|x\| = \sup_{\alpha \in D}|x(\alpha)|$.

For each positive integer k, let c_k stand for the discrete hypercube $\{u \in D : |\alpha_j| \le k \; \forall j\}$ (α_j is the jth component of α), and let $\ell_1(D)$ denote the set of $\mathbb{R}$-valued maps g on D such that

$$\sup_k \sum_{\beta \in c_k} |g(\beta)| < \infty$$

For each $g \in \ell_1(D)$ the sum $\Sigma_{\beta \in c_k} g(\beta)$ converges to a finite limit as $k \to \infty$, and we denote this limit by $\Sigma_{\beta \in D} g(\beta)$.

Define maps Q_k and E_k from $\ell_\infty(D)$ into itself by $(Q_k x)(\alpha) = x(\alpha)$, $\alpha \in c_k$ and $(Q_k x)(\alpha) = 0$ otherwise, and $(E_k x)(\alpha) = x(\alpha)$, $\alpha \notin c_k$ and $(E_k x)(\alpha) = 0$ otherwise.

In the next section H stands for any linear map from $\ell_\infty(D)$ into itself that satisfies the condition that

$$\sup_k (HQ_ku)(\alpha) < \infty \tag{2.24}$$

for each $u \in \ell_\infty(D)$ and each $\alpha \in D$. This condition is met whenever H is continuous because then $(HQ_ku)(\alpha) \leq |(HQ_ku)(\alpha)| \leq \|HQ_ku\| \leq \|H\| \cdot \|Q_ku\| \leq \|H\| \cdot \|u\|$.[52]

Our Theorem In the following theorem, $h(\cdot,\beta)$ for each $\beta \in D$ is defined by $h(\cdot,\beta) = H\delta_\beta$, where $(\delta_\beta)(\alpha) = 1$ for $\alpha = \beta$ and $(\delta_\beta)(\alpha)$ is zero otherwise. Of course $h(\cdot,\beta)$ is the response of H to a unit "impulse" occurring at $\alpha = \beta$.

Theorem. For any H as described, and for each $\alpha \in D$ and each $x \in \ell_\infty(D)$,

  (i) g defined on D by $g(\beta) = h(\alpha,\beta)x(\beta)$ belongs to $\ell_1(D)$.
  (ii) $\lim_{k\to\infty}(HE_kx)(\alpha)$ exists and is finite.
  (iii) We have

$$(Hx)(\alpha) = \sum_{\beta \in D} h(\alpha,\beta)x(\beta) + \lim_{k\to\infty}(HE_kx)(\alpha)$$

Proof. Let $\alpha \in D$ and $x \in \ell_\infty(D)$ be given. By the linearity of H and the definition of Q_k,

$$\sum_{\beta \in c_k} h(\alpha,\beta)u(\beta) = H\left(\sum_{\beta \in c_k} u(\beta)\delta_\beta\right)(\alpha) = H(Q_ku)(\alpha) \tag{2.25}$$

for each k and each $u \in \ell_\infty(D)$. In particular, for u given by $u(\beta) = \text{sgn}[h(\alpha,\beta)]\ \forall\beta$,

$$\sum_{\beta \in c_k} |h(\alpha,\beta)| = H(Q_ku)(\alpha)$$

for each k. Thus, by (2.24), $h(\alpha,\cdot)$ belongs to $\ell_1(D)$ and so does g of (i) of the theorem. By (i) the extreme left and right sides of (2.25) with $u = x$ converge as $k \to \infty$, and one has

---

[52] Here we have used the well-known fact that boundedness and continuity are equivalent for a liner operator between normed linear spaces.

$$\sum_{\beta \in D} h(\alpha, \beta) x(\beta) = \lim_{k \to \infty} H(Q_k x)(\alpha) \qquad (2.26)$$

Since

$$H(E_k x)(\alpha) = (Hx)(\alpha) - H(Q_k x)(\alpha)$$

for each k, it is clear that (ii) holds. Finally,

$$\lim_{k \to \infty} H(E_k x)(\alpha) = (Hx)(\alpha) - \lim_{k \to \infty} H(Q_k x)(\alpha)$$

which, together with (2.26), completes the proof.

We note that for $D = \mathbf{Z}_+^m$ and H causal in the usual sense (see Section G.2), the term $\lim_{k \to \infty} (HE_k x)(\alpha)$ is always zero. In the next section an extension result for shift-invariant maps defined on $\ell_\infty(\mathbf{Z}^m)$ is given from which it follows that there are maps H that do not have a generalized convolution representation (i.e., that there are H's for which the additional term is not always zero).

Using the theorem just proved, it is not difficult to give other criteria under which H has a generalized convolution representation. This is illustrated by the following.

Proposition. Under the hypotheses of our theorem, the following five statements are equivalent.

(i) For each $\alpha \in D$ and each $x \in \ell_\infty(D)$, g defined on D by $g(\beta) = h(\alpha,\beta)x(\beta)$ belongs to $\ell_1(D)$ and we have $(Hx)(\alpha) = \Sigma_{\beta \in D} h(\alpha,\beta)x(\beta)$.

(ii) $\lim_{k \to \infty}(HE_x x)(\alpha) = 0$ for $\alpha \in D$ and $x \in \ell_\infty(D)$.

(iii) For each $\alpha \in D$, $\lim_{k \to \infty}(HE_k x)(\alpha) = 0$ uniformly for $x \in \ell_\infty(D)$ with $\|x\| \le 1$.

(iv) For each $\alpha \in D$, $\lim_{k \to \infty}(HE_{\alpha,k} x)(\alpha) = 0$ uniformly for $x \in \ell_\infty(D)$ with $\|x\| \le 1$, where $(E_{\alpha,k} x)(\beta) = x(\beta)$ for $(\beta - \alpha) \notin c_k$ and $(E_{\alpha,k} x)(\alpha) = 0$ otherwise.

(v) For each $\alpha \in D$ there is an $\mathbb{R}$-valued function q defined on D, with q belonging to $\ell_1(D)$, such that $|(Hx)(\alpha)| \le \Sigma_{\beta \in D} |q(\beta)x(\beta)|$ for $x \in \ell_\infty(D)$.

The proof of this proposition is very direct: We see that (i) $\Rightarrow$ (v) [take $q(\beta) = |h(\alpha,\beta)|$], and that (v) $\Rightarrow$ (iv) [use the fact that

$\Sigma_{(\beta-\alpha)\notin c_k}|q(\beta)| \to 0$ as $k \to \infty] \Rightarrow$ (iii) [given $\varepsilon > 0$, choose k_ε so that $|(HE_{n,\alpha}x)(\alpha)| < \varepsilon$ for $k > k_\varepsilon$ and x as indicated, and observe that $|(HE_k x)(\alpha)| < \varepsilon$ for $k > \max_j|\alpha_j| + k_\varepsilon$ and x as described] $\Rightarrow$ (ii) $\Rightarrow$ (i).

G.2 An Extension Proposition

We begin with some additional preliminaries: Let M denote a linear manifold in $\ell_\infty(\mathbf{Z}^m)$ that is closed under translation in the sense that $T_\beta M = M$ for each $\beta \in \mathbf{Z}^m$, where T_β is the usual shift map defined on M for each $\beta \in \mathbf{Z}^m$ by $(T_\beta x)(\alpha) = x(\alpha - \beta), \alpha \in \mathbf{Z}^m$. Assume also that $y \in M$ implies that $z \in M$, where $z(\alpha) = y(\alpha)$ for $\alpha_j \leq 0 \; \forall j$, and $z(\alpha) = 0$ otherwise. We do not rule out the possibility that $M = \ell_\infty(\mathbf{Z}^m)$.

Let A be a linear map of M into $\ell_\infty(\mathbf{Z}^m)$. Such an A is *shift invariant* if

$$(Ax)(\alpha - \beta) = (AT_\beta x)(\alpha) \qquad \alpha \in \mathbf{Z}^m$$

for each $\beta \in \mathbf{Z}^m$ and $x \in M$. The map A is *causal* if

$$x(\alpha) = y(\alpha) \text{ whenever } \alpha_j \leq \beta_j \forall j \Rightarrow (Ax)(\beta) = (Ay)(\beta)$$

for each $\beta \in \mathbf{Z}^m$ and every x and y in M. It is bounded if $\|A\|_M :=$ $\sup\{\|Ax\| : x \in M, \|x\| \leq 1\} < \infty$, in which $\|\cdot\|$ is the norm in $\ell_\infty(\mathbf{Z}^m)$. Our result is the following.

Proposition. Let A be shift invariant and bounded. Then there exists a bounded linear shift-invariant map B from $\ell_\infty(\mathbf{Z}^m)$ into itself that extends A in the sense that B is causal if A is causal and $Bx = Ax, x \in M$.

Proof. By the shift invariance of A, we have $(Ax)(\alpha) = (AT_{-\alpha}x)(0)$ for all α and all $x \in M$. The map $(A\cdot)(0)$ is a bounded linear functional on M, because

$$|(Ay)(0)| = |(AT_{-\alpha}T_\alpha y)(0)| = |(AT_\alpha y)(\alpha)|$$
$$\leq \sup_\beta |(AT_\alpha y)(\beta)| \leq \|A\|_M \cdot \|T_\alpha y\| = \|A\|_M \cdot \|y\|$$

for $y \in M$. When A is causal, $(A\cdot)(0)$ has the property that $(Ay)(0) = 0$ for any $y \in M$ for which $y(\alpha) = 0$ for $\alpha_j \leq 0 \; \forall j$. By the

Hahn–Banach theorem (Bachman and Narici 1966, p. 178)[53] there is a bounded linear functional $\mathcal{F}$ that extends $(A \cdot)(0)$ to all of $\ell_\infty(\mathbf{Z}^m)$. Set $\mathcal{G} = \mathcal{F}$ if A is not causal, and if A is causal define $\mathcal{G}$ on $\ell_\infty(\mathbf{Z}^m)$ by $\mathcal{G}y = \mathcal{F}Py$ where P is the linear operator given by $(Py)(\alpha) = y(\alpha)$, $\alpha_j \leq 0 \; \forall_j$ and $(Py)(\alpha) = 0$ for $\alpha_j > 0$ for some j. Define B on $\ell_\infty(\mathbf{Z}^m)$ by $(Bx)(\alpha) = \mathcal{G}T_{-\alpha}x$. It is easy to check that B is a linear shift-invariant bounded map into $\ell_\infty(\mathbf{Z}^m)$, that B extends A to $\ell_\infty(\mathbf{Z}^m)$, and that B is causal if A is causal.[54] This completes the proof.

Since the set L of elements x of $\ell_\infty(\mathbf{Z}^m)$ such that $x(\alpha)$ approaches a limit as $\max_j\{\alpha_j\} \to -\infty$ is a linear manifold that is closed under translation, and since

$$(Ax)(\alpha) = \lim_{\max_j\{\beta_j\} \to -\infty} x(\beta)$$

defines a shift-invariant bounded causal linear map of L into $\ell_\infty(\mathbf{Z}^m)$, it follows from our proposition that there exist maps H, even causal time-invariant maps H, of the kind addressed by our theorem for which the term $\lim_{k \to \infty}(HE_kx)(\alpha)$ is not always zero. More explicitly, the associate B via our proposition of the A just described satisfies $\lim_{k \to \infty}(BE_kx)(\alpha) = \lim_{\max_j\{\beta_j\} \to -\infty} x(\beta)$ for $x \in L$. And an example of an H of the type addressed by the theorem for which the additional term is not always zero and H is not shift invariant is obtained by adding to this B any linear bounded map of $\ell_\infty(\mathbf{Z}^m)$ into itself that is not shift invariant and has a representation without an additional term.

A proposition similar to the one above can be given to show that there are bounded linear continuous-time time-invariant input–output maps that do not possess certain convolution representations.[55]

---

[53] The Hahn–Banach theorem tells us that a bounded linear functional on a subspace of a normed linear space can be extended to the entire space with preservation of norm.

[54] It is also true that B can be chosen so that it preserves the norm of A, in the sense that $\|B\| = \|A\|_M$.

[55] As suggested above, results along the lines of the theorem in this appendix hold also in continuous-time settings. More specifically (Sandberg 1998a), for continuous single-input–single-output causal input–output maps H that take bounded Lebesgue measurable inputs defined on $\mathbb{R}$ into bounded outputs on $\mathbb{R}$ such that a certain often-satisfied condition is met, $(Hx)(t) = \int_{-\infty}^{t} h(t, \tau)x(\tau)d\tau + \lim_{a \to -\infty}(HP_a x)(t)$ for all t, in which h has the usual continuous-time impulse-response interpretation and

Referring now to Section 2.3.2, the map B described above provides an example of a linear G that is myopic for no w. To see this, let S be the unit open ball in $l_\infty(\mathbf{Z}^m)$ centered at the origin and let B be as described. With w arbitrary, suppose that B is such that given some $\varepsilon \in (0,\frac{1}{2})$ there is a $\delta > 0$ with the property that $x \in S$ and $\sup_{\alpha \in \mathbf{Z}^m}\|w(\alpha)x(\alpha)\| < \delta$ imply $|(Bx)(0)| < \varepsilon$. Let y be any element of $S \cap L$ such that $y(\beta)$ approaches $\frac{3}{4}$ as $\max_j\{\beta_j\} \to -\infty$. Let z denote $E_k y$, with k chosen so that $\sup_{\alpha \in \mathbf{Z}^m}\|w(\alpha)z(\alpha)\| < \delta$. Here $z \in S$ and $\sup_{\alpha \in \mathbf{Z}^m}\|w(\alpha)z(\alpha)\| < \delta$ is satisfied, but $|(Bx)(0)| = \frac{3}{4} > \varepsilon$.[56]

Appendix H: Notes on Uniform Approximation of Shift-Varying Systems

It is shown in Sandberg (1998) that the elements K of a large class of continuous-time causal input–output maps can be uniformly approximated arbitrarily well using a certain structure if and only if K is uniformly continuous. For the case addressed, the system inputs and outputs are defined on a finite interval $[0, t_f]$. In this appendix we give corresponding results for the case in which K is not necessarily causal and inputs and outputs are defined on a discrete set $\{0, 1, \ldots, a_1\} \times \ldots \times \{0, 1, \ldots, a_m\}$, in which $a_1, \ldots, a_m$ are positive integers. As in Section 2.3.1 our approximating structure involves certain functions that can be chosen in different ways. When these functions are taken to be certain polynomial functions, the input–output map of the structure is a generalized shift-varying Volterra series.

H.1 Preliminaries and the Approximation Result With
$a_1, \ldots, a_m$ any positive integers, and with $A = \{0, 1, \ldots, a_1\} \times \cdots \times \{0, 1, \ldots, a_m\}$, let $L(A)$ denote the linear space of maps x from A to

---

$(P_a x)(\tau) = x(\tau)$ for $\tau \le a$ and $(P_a x)(\tau) = 0$ otherwise. An example is given in Sandberg (1998a) of an H for which $\lim_{a \to \infty}(HP_a x)(t)$ is not always zero. Corresponding results for noncausal input–output maps can be proved too. In this case the extra term involves also the behavior of the input for large values of its argument, as it does in the discrete-time case addressed by our theorem. And corresponding results in the setting of inputs and outputs of a finite number of variables have been proved (Sandberg 1999b) starting with the approach described in Ball and Sandberg (1990). The theorem in this section was given for the first time in (Sandberg 1997). A more general version can be found in Sandberg (1998b).

[56] A similar example can be given of a linear G in the continuous-time setting of Sandberg and Xu (1997a) that is myopic for no w.

$\mathbb{R}^n$. For each $\alpha \in A$, let P_α denote the map from $L(A)$ into itself defined by $(P_\alpha x)(\beta) = x(\beta)$, $\beta_j \leq \alpha_j$ $\forall j$ and $(P_\alpha x)(\beta) = 0$ otherwise.

We view $L(A)$ as a normed space with some norm $\|\cdot\|_{L(A)}$ that is regular in the sense that $\|P_\alpha x\|_{L(A)} \leq \|x\|_{L(A)}$ for all α and x.[57] The norm $\|\cdot\|_{L(A)}$ can be chosen in many ways. One choice is given by

$$\|x\|_{L(A)} = \max_{\alpha \in A} \|x(\alpha)\|$$

Let $M(A)$ stand for the space of real-valued maps y defined on A, with some norm $\|y\|_{M(A)}$. It will become clear that the choice of this norm is unimportant because all norms on finite dimensional linear spaces are equivalent.

With c any positive number, let E denote $\{x \in L(A): \|x\|_{L(A)} \leq c\}$. Let K map E to $M(A)$. K is the input–output map that we wish to approximate. Of course, the set E is the set of inputs corresponding to the map K. By K is *causal* we mean that for each $\alpha \in A$ one has $(Kx)(\alpha) = (Ky)(\alpha)$ for x and y in E with $x(\beta) = y(\beta)$, $\beta_j \leq \alpha_j$ $\forall j$.

In the theorem below, which is the main result in this appendix, ℓ denotes the dimension of $M(A)$ [of course, $l = (1 + a_1) \times \cdots \times (1 + a_m)$, but this fact is not used in what follows], $v_1, \ldots, v_\ell$ is any set of basis elements for $M(A)$, and $\mathcal{H}$ stands for the set of maps H from E to $\mathbb{R}$ such that for each H there are an $n_0 \in \mathbb{N}$, maps $g_1, \ldots, g_{n_0}$ from A to $\mathbb{R}^n$, and an $N \in D_{n_0}$ such that $Hx = N[y_1(x), \ldots, y_{n_0}(x)]$ where each y_j is defined on E by

$$y_j(x) = \sum_{\alpha \in A} \langle g_j(\alpha), x(\alpha) \rangle$$

Theorem. The following two statements are equivalent.

(i) K is continuous (i.e., continuous with respect to the norms $\|\cdot\|_{L(A)}$ and $\|\cdot\|_{M(A)}$).

(ii) For each $\varepsilon > 0$, there are elements $H_1, \ldots, H_\ell$ of $\mathcal{H}$ such that

$$\left| (Kx)(\alpha) - \sum_{j=1}^{\ell} v_j(\alpha)(H_j x) \right| < \varepsilon \qquad \alpha \in A \qquad (2.27)$$

for all $x \in E$.

---

[57] As is well known, norms are typically regular. However, it is not difficult to give an example of a norm that is not regular.

In addition, $H_j x$ in (ii) can be replaced with $H_j P_\alpha x$ if K is causal.

Proof. We use two lemmas. More specifically, using the hypothesis that E is compact in $L(A)$, the following can be proved using direct modifications of material in Appendix D.

Lemma 5. Let $R: E \to \mathbb{R}$ be continuous with respect to $\|\cdot\|_{L(A)}$. Then for each $\varepsilon > 0$, there is an $H \in \mathcal{H}$ such that $|Rx - Hx| < \varepsilon, x \in E$.

In order to describe our second lemma, let Q be any set of $\mathbb{R}$-valued continuous maps defined on E that is dense in the set $U(E)$ of continuous real functionals on E, in the sense that for each $R \in U(E)$ and each $\varepsilon > 0$ there is a $Q \in Q$ such that $\max_{x \in E} |R(x) - Q(x)| < \varepsilon$. Let $\mathcal{P}$ denote the set of all maps from E to $M(A)$ of the form $\sum_{j=1}^{\ell} Q_j(\cdot) v_j$ where the Q_j belong to Q. Our second lemma,[58] which is proved in Section H.3 is the following.

Lemma 6. Let F be a map from E to $M(A)$. Then F is continuous if and only if for each $\varepsilon > 0$ there exists a $P \in \mathcal{P}$ such that $\|Fx - Px\|_{M(A)} < \varepsilon$ for $x \in E$.

Continuing with the proof of the theorem, suppose that (i) is satisfied. Using Lemmas 5 and 6, we see that for each $\varepsilon > 0$ there are elements $H_1, \dots, H_\ell$ of $\mathcal{H}$ such that

$$\left| (Ky)(\alpha) - \sum_{j=1}^{\ell} v_j(\alpha)(H_j y) \right| < \varepsilon \qquad \alpha \in A$$

for all $y \in E$. This is (ii). And if K is causal, (ii) holds with $H_j x$ replaced with $H_j P_\alpha x$ because $P_\alpha x \in E$ whenever $x \in E$ and $\alpha \in A$, and K causal implies that $(Kx)(\alpha) = (KP_\alpha x)(\alpha)$ for all x and α.

Now assume that either (ii) holds or that (ii) holds with $H_j x$ replaced with $H_j P_\alpha x$. Since each v_j belongs to $M(A)$, and since the maps J_j defined on E by either $J_j x = H_j x$ or $(J_j x)(\alpha) = H_j P_\alpha x$ are continuous from E to $\mathbb{R}$ or from E to $M(A)$, respectively,[59] it follows that

---

[58] This lemma is a simple version of a result that plays a central role in Dingankar and Sandberg (1995) [see also Sandberg (1998)]. The proof given in this appendix is much simpler than the corresponding proofs in the references cited.

[59] The continuity of the J_j is a consequence of the boundedness of the set E, the Schwarz inequality in $\mathbb{R}^n$, and the fact that the elements of D_{n_0} are uniformly continuous on compact sets.

K is the uniform limit of a sequence of continuous maps. And from this it follows that K is continuous. This completes the proof of the theorem.

H.2 Special Case: Finite Volterra-Like Series Approximations

Suppose that K is causal. If the D_{n_0} are chosen to be certain sets of polynomial functions, the approximating maps

$$\sum_{j=1}^{\ell} v_j(\alpha)(H_j P_\alpha \cdot)$$

of our theorem are generalizations of finite discrete shift-varying Volterra series. More specifically, suppose that the maps f of each D_{n_0} are such that $f(w)$ is a real polynomial in the components of w, and consider the $n = 1$ case. By writing products of sums as iterated sums, it is not difficult to see that $\sum_{j=1}^{\ell} v_j(\alpha)(H_j P_\alpha x)$ has the form

$$k_0(\alpha) + \sum_{j=1}^{m_1} \sum_{\beta_j \in A_\alpha} \cdots \sum_{\beta_1 \in A_\alpha} k_j(\alpha, \beta_1, \ldots, \beta_j) x(\beta_1) \cdots x(\beta_j) \quad (2.28)$$

in which n_1 is a positive integer, $A_\alpha = \{\beta \in A : \beta_j \le \alpha_j \; \forall j\}$, and each k_j $(j \ge 0)$ is a real-valued map defined on $A^{(j+1)}$.[60] On the other hand, any map $V : E \to M(A)$ defined by the condition that $(Vx)(\alpha)$ is given by (2.28) is continuous.[61] We have therefore proved the following concerning causal K's (see the last part of the proof of our theorem). For $n = 1$ and with $\mathcal{V}$ the set of all maps $V : E \to M(A)$ such that $(Vx)(\alpha)$ is given by (2.28) for some $n_1 \in \mathbb{N}$ and some $k_0, \ldots, k_{n_1}$, (i) of the theorem holds (i.e., K is continuous) if and only if for any $\varepsilon > 0$ there is a $V \in \mathcal{V}$ such that $|(Kx)(\alpha) - (Vx)(\alpha)| < \varepsilon$ for $x \in E$ and $\alpha \in A$.

In fact, it is not difficult to check that this proposition holds for all $n \ge 1$ with the understanding that each k_j in (2.28) is row n^j-vector valued, and that $x(\beta_1) \ldots x(\beta_j)$ in (2.28) denotes the corresponding Kronecker product (which is column n^j-vector valued, assuming of course that we view the elements of $\mathbb{R}^n$ as column vectors).

---

[60] Of course, for $m = 1$, (2.28) takes the more familiar form $k_0(\alpha) + \sum_{j=1}^{n_1} \sum_{\beta_j=0}^{\alpha} \ldots \sum_{\beta_1=0}^{\alpha} k_j(\alpha, \beta_1, \ldots, \beta_j) x(\beta_1) \cdots x(\beta_j)$.

[61] The continuity of V is a consequence of the boundedness of the set E and the identity $c_1 c_2 \cdots c_j - b_1 b_2 \cdots b_j = c_1 c_2 \cdots c_{(j-1)}(c_j - b_j) + c_1 \cdots c_{(j-2)}(c_{(j-1)} - b_{(j-1)}) b_j + \ldots + (c_1 - b_1) b_2 \cdots b_j$ for real numbers $c_1, \ldots, c_j, b_1, \ldots, b_j$.

A similar proposition holds for K's that are not necessarily causal. In this case (2.28) is replaced with

$$k_0(\alpha) + \sum_{j=1}^{m_1} \sum_{\beta_j \in A} \cdots \sum_{\beta_1 \in A} k_j(\alpha, \beta_1, \ldots, \beta_j) x(\beta_1) \cdots x(\beta_j) \qquad (2.29)$$

and K is continuous if and only if $(Kx)(\alpha)$ can be approximated arbitrarily well in the sense described above by a sum of the form (2.29).

H.3 Proof of Lemma 6 Let continuous F from E to $M(A)$ be given, and let $x \in E$ and any $\varepsilon > 0$ also be given. Since the v_j form a basis for $M(A)$, we have $Fx = \sum_{j=1}^{\ell} f_j(Fx)v_j$ for $x \in E$, where the f_j are continuous linear functionals. Using the fact that the $f_j(F\cdot)$ are continuous, choose $Q_1, \ldots, Q_l \in Q$ so that

$$\max_j \max_{x \in E} \left(|f_j(Fx) - Q_j x| \cdot \|v_j\|_{M(A)} \right) < \varepsilon / \ell$$

Then set $P = \sum_{j=1}^{\ell} Q_j(\cdot)v_j$. This gives

$$\|Fx - Px\|_{M(A)} \le \sum_{j=1}^{\ell} |f_j(Fx) - Q_j x| \cdot \|v_j\|_{M(A)} < \varepsilon$$

for $x \in E$.

On the other hand, if F can be approximated as indicated in the lemma, then F is the uniform limit of a sequence of continuous maps, and F is therefore continuous. This proves the lemma.

Appendix I Approximation with the Doubly Finite Volterra Series of Section 2.4.3

As in Section 2.4.3, let $\mathcal{M}$ stand for the set of all causal time-invariant maps from S to the real-valued functions defined on $\mathbf{Z}_+$, and let $\mathcal{V}$ denote the family of all members of $\mathcal{M}$ that have a representation of the form

$$\sum_{j=1}^{p} \sum_{n_j=0}^{n} \cdots \sum_{n_1=0}^{n} k_j(n - n_1, \ldots, n - n_j) s(n_1) \cdots s(n_j)$$

for some p and some $k_1, \ldots, k_p$ with each $k_j(n_1, \ldots, n_j) = 0$ when one or more of the $n_1, \ldots, n_j$ exceed some number α. In Section 2.4.3 the "if part" of the following theorem is proved, and here we prove the converse.

Theorem 9. A map $J \in \mathcal{M}$ can be uniformly approximated arbitrarily well by an element of $\mathcal{V}$ if and only if J has approximately finite memory and the functionals $J(\cdot)(n)$ are continuous with respect to the metric on S given by $\rho(x, y) = \sup\{\|x(j) - y(j)\| : j \in \mathbf{Z}_+\}$.

Proof of the "Only if Part." Assume that $J \in \mathcal{M}$ can be uniformly approximated arbitrarily well by an element of $\mathcal{V}$. Let $\varepsilon > 0$ be given. Choose $v \in \mathcal{V}$ such that $|(Js)(n) - (vs)(n)| < \varepsilon/2$ for all n. Let $\alpha \in \mathbf{Z}_+$ satisfy $(vW_{n,\alpha}s)(n) = (vs)(n)$ for all n and s (there is such an α because of the properties of the k_j). Observe that for all n and s

$$|(Js)(n) - (JW_{n,\alpha}s)(n)| \leq |(Js)(n) - (vs)(n)|$$
$$+ |(vW_{n,\alpha}s)(n) - (JW_{n,\alpha}s)(n)| < \varepsilon$$

showing that J has approximately finite memory.[62]

Now select any n and any element of $\mathcal{V}$, and consider the corresponding functional v_n defined on S_n by

$$v_n(u) = \sum_{j=1}^{p} \sum_{n_j=0}^{n} \cdots \sum_{n_1=0}^{n} k_j(n - n_1, \ldots, n - n_j) u(n_1) \cdots u(n_j)$$

Since this functional is continuous it follows (see the corresponding part of the proof of Theorem 1 and the related comment in Section 2.3 concerning the continuity of functionals) that the functionals $J(\cdot)(n)$ are continuous. This completes the proof.

A sufficient condition for the functionals $J(\cdot)(n)$ of Theorem 9 to be continuous is that J is a continuous map of S into the metric space of bounded real-valued functions on $\mathbf{Z}_+$, with the metric the natural extension of the metric on S. With $\mathcal{M}_\infty$ denoting such continuous elements of $\mathcal{M}$ we have the following useful result.

---

[62] This observation has the nice generalization that $J \in \mathcal{M}$ has approximately finite memory if and only if it can be uniformly approximated arbitrarily well by finite-memory maps in $\mathcal{M}$, where by a finite-memory map $M \in \mathcal{M}$ we mean that M satisfies $(MW_{n,\alpha}s)(n) = (Ms)(n)$ for all n and s and some α.

Corollary 2. $J \in \mathcal{M}_\infty$ can be uniformly approximated arbitrarily well by an element of $\mathcal{V}$ if and only if J has approximately finite memory.

Appendix J   Proof of Theorem 7

Suppose that (i) holds and choose any $n \in \mathbf{Z}_+$. Then, using the continuity of N, the continuity of the $q(l,n,\cdot)$, and the continuity of W_β^i, since the real-valued function J_n can be uniformly approximated arbitrarily well over U_n by continuous functions, J_n is continuous.

In addition, for each $\varepsilon > 0$ there are r, k, elements $l_1, \ldots, l_k$ of L, and an $N \in D_k$ such that

$$|J_n v - N(Qv)| < \varepsilon$$

for all $v \in U_n$, where $Qv = [q(l_1',n,v), \ldots, q(l_k',n,v)]$ and the l_j' arc elements of L that satisfy $(l_j'u)(\gamma) = (l_j W_\gamma^r u)(\gamma)$ for each j. Thus, by Lemma 4, A.2 holds.

Assume now that (ii) is met, and let $\varepsilon > 0$ be given. Choose $r \in \mathbf{Z}_+$ so that

$$|(Ku)(\gamma) - (KW_\gamma^r u)(\gamma)| < \varepsilon/2 \qquad \gamma \in \mathbf{Z}^m \qquad (2.30)$$

for $u \in U$.

By the continuity of J_r and the compactness of U_r, and by our lemma (in Section 2.4.2), there are k, elements $l_1, \ldots, l_k$ of L, and an $N \in D_k$ such that

$$|J_r v - N(Pv)| < \varepsilon/2 \qquad (2.31)$$

for all $v \in U_r$, where $Pv = [q(l_1,r,v), \ldots, q(l_k,r,v)]$. Now let $u \in U$ and $\gamma \in \mathbf{Z}^m$ be given.

By the shift invariance of K and the l_j,

$$(KW_\gamma^r u)(\gamma) - N[(MW_\gamma^r u)(\gamma)] = J_r v - N(Pv)$$

with M as described in the theorem and $v(\beta) = u(\beta + \gamma)$ for $\beta \in c_r$. By (2.31)

$$|(KW_\gamma^r u)(\gamma) - N[(MW_\gamma^r u)(\gamma)]| < \varepsilon/2$$

This together with (2.30) gives

$$|(Ku)(\gamma) - N[(MW_\gamma^r u)(\gamma)]| < \varepsilon$$

which completes the proof of the theorem.

Appendix K   Multivariable Volterra Series Approximations

Assume that $L = L_0$ with the ϕ's linear. It is easy to check that the separation condition of Theorem 8 is met (see the corresponding observation in Section 2.4.3.). Assume also that the D_k are linearring maps. Let Q denote the approximating map of Theorem 5 given by $(Qu)(\gamma) = N[(MW_\gamma^r u)](\gamma)]$, and let $l_1', \ldots, l_k'$ be elements of L such that $(l_j'u)(\gamma) = (l_j W_\gamma^r u))(\gamma)$ Consider the $d = 1$ case. By writing products of sums as iterated sums, one can see that $(Qu)(\gamma)$ has the form

$$\sum_{j=1}^{p} \sum_{\beta_j \in \mathbf{Z}^m} \cdots \sum_{\beta_1 \in \mathbf{Z}^m} k_j(\gamma - \beta_1, \ldots, \gamma - \beta_j) u(\beta_1) \cdots u(\beta_j) \qquad (2.32)$$

in which p is a positive integer, and each $k_j(\gamma - \beta_1, \ldots, \gamma - \beta_j)$ is a finite linear combination of products of the form $a'_{i_j(1)}(\gamma - \beta_1) a'_{i_j(2)} (\gamma - \beta_2) \cdots a'_{i_j(j)}(\gamma - \beta_j)$, where the indices $i_j(1), \ldots, i_j(j)$ are drawn from $\{1, \ldots, k\}$ and each a_i' is the kernel (i.e., discrete impulse response) associated with l_i'. Note that each k_j vanishes whenever the magnitude of any one of the components of its j arguments exceeds r.

Expression (2.32) is a (multivariable) doubly finite Volterra series approximant for K. Let $\mathcal{M}_m$ stand for the set of all shift-invariant maps from U to the real-valued functions defined on $\mathbf{Z}^m$, and let $\mathcal{V}_m$ denote the family of all members of $\mathcal{M}_m$ that have a representation of the form (2.32) for some p and some $k_1, \ldots, k_p$ such that for each j we have $k_j(\beta_1, \ldots, \beta_j) = 0$ when one or more of the components of some $\beta_1, \ldots, \beta_j$ have magnitudes that exceed some number r. We have seen that sufficient conditions for a member of $\mathcal{M}_m$ to be uniformly approximated arbitrarily well by an element of $\mathcal{V}_m$ are that the member is myopic and that its corresponding J_n functionals are continuous. A direct modification of the proof described in Appendix I shows that these conditions are necessary. Also, a corollary of the kind described in Appendix I holds here too, showing that any member of a large class of continuous multivariable shift invariant maps[63] has arbitrarily good doubly finite Volterra approximants if and only if it is myopic.

Special Symbols

$A \times A$	Cartesian product of set A with itself
$a \vee b, a \wedge b$	lattice operations
$C_1, \ldots, C_m$	sets of signals
$C(\mathbb{R}^{n_0}, \mathbb{R})$	set of continuous maps from $\mathbb{R}^{n_0}$ to $\mathbb{R}$
$cl(QU)$	set closure of QU in X_w
$\exp(\cdot)$	exponential function
$f: C \to \mathbb{R}$	f is a map from C to $\mathbb{R}$
G	input–output map from S to the set of $\mathbb{R}$-valued functions on $\mathbf{Z}^m$
$(Gs)(n)$	G evaluated at $s \in S$ evaluated at n (S is the input set)
$L_2(a, b)$	linear space of real-valued square integrable functions on $[a,b]$
$\mathrm{Lip}(\kappa)$	set of Lipschitz continuous elements, Lipschitz constant κ
$L_p(\mathbb{R}^n)$	linear space of real-valued pth-power integrable functions
$\max\{0, c_1, \ldots, c_p\}$	maximum of $0, c_1, \ldots, c_p$
$\min\{0, c_1, \ldots, c_p\}$	minimum of $0, c_1, \ldots, c_p$
N	nonlinear static map
$\mathbb{N}$	set of positive integers
$\mathbb{R}$	set of real numbers
$\mathbb{R}^n$	set of real n-vectors
$\mathbb{R}_+$	set of nonnegative real numbers
$\sup_{\alpha \in A} B(\alpha)$	least upper bound of B over A
T_β	shift operator
$\cup_j B_j$	union over j of B_j
$\langle u, v \rangle$	inner product of u and v in $\mathbb{R}^n$
$(W_{\beta,\alpha}s)(n) = \begin{cases} s(n) \\ 0 \end{cases}$	$\begin{array}{l} \beta - \alpha \leq n \leq \beta \\ \text{otherwise} \end{array}$

---

[63] Specifically, these maps take U into the metric space of bounded functions on $\mathbf{Z}^m$, with the metric the natural extension of the metric on U.

X	real normed linear space
X^*	set of real bounded linear functionals on X
$X(\mathbf{Z}^m, \mathbb{R}^n)$	set of all $\mathbb{R}^n$-valued maps defined on $\mathbf{Z}^m$
$\|x\|$	norm of x
(x_1, x_2)	ordered pair
$\{x \in A : f(x) \neq 0\}$	set of $x \in A$ for which $f(x) \neq 0$
$\mathbf{Z}$	set of all integers
$\mathbf{Z}_+$	set of nonnegative integers
$\mathbf{Z}_-$	set of nonpositive integers
$[0, 1]^n$	n-dimensional interval
ε	positive constant
$|\mu|(I)$	total variation of μ on I
ρ, ρ_n	metrics (ρ also used as a positive constant)
Σ_{β_j}	sum over β_j
$\Sigma_{j=1}^{p}$	sum from 1 to ρ

REFERENCES

Bachman, G., and L. Narici, 1966, *Functional Analysis* (New York: Academic Press).

Ball, D., and I. W. Sandberg, July 1990, "*g*- and *h*-representations for nonlinear maps," *J. Math. Anal. Applic.*, vol. 149, no. 2.

Borodziewicz, W. J., K. J. Jaszczak, and M. A. Kowalski, 1983, "A note on mathematical formulation of discrete-time linear systems," *Signal Processing*, vol. 5, pp. 369–375.

Boyd, S., and L. O. Chua, Nov. 1985, "Fading memory and the problem of approximating nonlinear operators with Volterra series," *IEEE Trans. Circuits Syst.*, vol. CAS-32, no. 11, pp. 1150–1161.

Chen, T., and H. Chen, Nov. 1993, "Approximations of continuous functionals by neural networks with application to dynamical systems," *IEEE Trans. Neural Networks*, vol. 4, no. 6, pp. 910–918.

Chen, T., and H. Chen, July 1995, "Universal approximation to nonlinear operators by neural networks with arbitrary activation functions and its applications to dynamic systems," *IEEE Trans. Neural Networks*, vol. 6, no. 4, pp. 911–917.

Coleman, B. D., and V. J. Mizel, 1966, "Norms and semi-groups in the theory of fading memory," *Arch. Rational Mech. Anal.*, vol. 23, pp. 87–123.

Coleman, B. D., and V. J. Mizel, 1968, "On the general theory of fading memory," *Arch. Rational Mech. Anal.*, vol. 6, pp. 180–231.

Cybenko, G., 1989, "Approximation by superposition of a single function," *Math. Control, Signals Syst.*, vol. 2, pp. 303–314.

De Vries, B., J. C. Principe, and P. Guedes de Oliveira, 1991, "Adaline with adaptive recursive memory," *Proc. IEEE-SP Workshop Neural Networks for Signal Processing,* pp. 101–110.

Dieudonné, J., 1969, *Foundations of Modern Analysis* (New York: Academic Press).

Dingankar, A., and I. W. Sandberg, Dec. 1995, "Network approximation of dynamical systems," *Proc. 1995 International Symposium on Nonlinear Theory and its Applications,* vol. 2, pp. 357–362, Las Vegas.

Edwards, R. E., 1995, *Functional Analysis* (New York: Dover).

Haykin, S., 1999, *Neural Networks,* 2nd ed. (Upper Saddle River, NJ: Prentice-Hall).

Hornik, K., 1991, "Approximation capabilities of multilayer feedforward networks," *Neural Networks,* vol. 4, pp. 251–257.

Kantorovich, L. V., and G. P. Akilov, 1982, *Functional Analysis* (Oxford: Pergamon).

Kishore, A. P., and J. B. Pearson, 1994, "Kernel representations and properties of discrete-time input-output systems," *Linear Algebra Its Applications,* Vol. 205–206, pp. 893–908.

Kolmogorov, A. N., and S. V. Fomin, 1970, *Introductory Real Analysis* (New York: Dover).

Liusternik, L. A., and V. J. Sobolev, 1961, *Elements of Functional Analysis* (New York: Frederick Ungar).

Mhaskar, H. N., and C. A. Micchelli, 1992, "Approximation by superposition of sigmoidal and radial basis functions," *Adv. App. Math.,* vol. 3, pp. 350–373.

Mukherjea, A., and K. Pothoven, 1978, *Real and Functional Analysis* (New York: Plenum Press).

Natanson, I., 1960, *Theory of Functions of a Real Variable,* Vol. II (New York: Frederick Ungar).

Oppenheim, A. V., A. S. Willsky, and S. H. Nawab, 1997, *Signals and Systems* (Upper Saddle River, NJ: Prentice Hall).

Park, J., and I. W. Sandberg, March 1993, "Approximation and radial-basis function networks," *Neural Computation,* vol. 5, no. 2, pp. 305–316.

Proakis, J. G., and D. G. Manolakis, 1992, *Digital Signal Processing* (New York: Macmillan).

Royden, H., 1968, *Real Analysis,* 2nd ed. (New York: MacMillan).

Rudin, W., 1976, *Principles of Mathematical Analysis,* 3rd ed. (New York: McGraw-Hill).

Sandberg, I. W., July 1983, "The mathematical foundations of associated expansions for mildly nonlinear systems," *IEEE Trans. Circuits Syst.,* vol. CAS-30, no. 7, pp. 441–455.

Sandberg, I. W., Jan. 1984, "A perspective on system theory," *IEEE Trans. Circuits Syst.*, vol. 31, no. 1, pp. 88–103.

Sandberg, I. W., 1991, "Structure theorems for nonlinear systems," *Multidimensional Systems and Signal Processing*, vol. 2, no. 3, pp. 267–286. (See also the Errata in vol. 3, no. 1, p. 101, 1992). A conference version of the paper appears in *Integral Methods in Science and Engineering-90* (Proceedings of the International Conference on Integral Methods in Science and Engineering, Arlington, Texas, May 15–18, 1990, ed. A. H. Haji-Sheikh) (New York: Hemisphere), pp. 92–110, 1991.

Sandberg, I. W., Jan. 1992a, "Approximations for nonlinear functionals," *IEEE Trans. Circuits Syst.*, vol. 39, no. 1, pp. 65–67.

Sandberg, I. W., July 1992b, "Approximately-finite memory and input-output maps," *IEEE Trans. Circuits Syst. I*, vol. 39, no. 7, pp. 549–556.

Sandberg, I. W., April 1993a, "A general structure for classification," *IEEE Trans. Circuits Syst.*, vol. 40, no. 4, pp. 288–289.

Sandberg, I. W., Oct. 1993b, "Uniform approximation and the circle criterion," *IEEE Trans. Automatic Control*, vol. 38, no. 10, pp. 1450–1458.

Sandberg, I. W., 1997, "Multidimensional nonlinear myopic maps, Volterra series, and uniform neural-network approximations," in D. Docampo, A. Figueiras-Vidal, and F. Perez-Gonzalez, eds., *Intelligent Methods in Signal Processing and Communications* (Boston: Birkhauser), pp. 99–128 (selected papers given at the Fourth Bayona Workshop on Intelligent Methods for Signal Processing and Communications, Bayona, Spain, June 1996).

Sandberg, I. W., Aug. 1998, "Notes on uniform approximation of time-varying systems on finite time intervals," *IEEE Trans. Circuits Syst. I*, vol. 45, no. 8, pp. 863–865. A conference version appears in Proceedings of the Fourth International Workshop on Nonlinear Dynamics of Electronic Systems, Seville, Spain, pp. 149–154, June 1996.

Sandberg, I. W., May 1998a, "A representation theorem for linear systems," *IEEE Trans. Circuits Syst. I*, vol. 45, no. 5, pp. 578–580.

Sandberg, I. W., 1998b, "A note on representation theorems for linear discrete-space systems," *J. Circuits, Syst., Signal Processing*, vol. 17, no. 6, pp. 703–707.

Sandberg, I. W., 1998c, "Separation conditions and criteria for uniform approximation of input-output maps," *Int. J. Circuit Theory Applic.*, vol. 26, pp. 243–252.

Sandberg, I. W., Jan. 1999a, "Bounds for discrete-time Volterra series representations," *IEEE Trans. Circuits Syst. I*, vol. 46, no. 1, pp. 135–139.

Sandberg, I. W., 1999b, "Multidimensional linear systems: the extra term," *Int. J. Circuit Theory Applic.*, vol. 27, pp. 415–420.

Sandberg, I. W., 2000, "Time-delay polynomial networks and quality of approximation," *IEEE Trans. Circuits Syst. I*, vol. 47, pp. 40–45.

Sandberg, I. W., and L. Xu, 1996, "Network approximation of input-output maps and functionals," *J. Circuits, Syst., Signal Processing*, vol. 15, no. 6, pp. 711–725.

Sandberg, I. W., and L. Xu, June 1997a, "Uniform approximation of multidimensional myopic maps," *IEEE Trans. Circuits Syst. I*, vol. 44, no. 6, pp. 472–485.

Sandberg, I. W., and L. Xu, 1997b, "Uniform approximation and gamma networks," *Neural Networks*, vol. 10, no. 5, pp. 781–784.

Sandberg, I. W., and L. Xu, April 1998, "Approximation of myopic systems whose inputs need not be continuous," *Multidimensional Syst. Signal Processing*, vol. 9, no. 2, pp. 207–225.

Schetzen, M., 1980, *The Volterra and Wiener Theories of Nonlinear Systems* (New York: Wiley).

Seibert, W. M., 1997, *Circuits, Signals, and Systems* (Cambridge, MA: MIT Press).

Shamma, J. S., and R. Zhao, 1993, "Fading-memory feedback systems and robust stability," *Automatica*, vol. 29, no. 1, pp. 191–200.

Stiles, B. W., I. W. Sandberg, and J. Ghosh, Nov. 1997, "Complete memory structures for approximating nonlinear discrete-time mappings," *IEEE Trans. Neural Networks*, vol. 8, no. 6, pp. 1397–1409.

Stone, M. H., March 1962, "A generalized Weierstrass approximation theorem," in *Studies in Modern Analysis*, R. C. Buck, ed., Vol. 1 of *MAA Studies in Mathematics* (Englewood Cliffs, NJ: Prentice-Hall), pp. 30–87.

Sutherland, W. A., 1975, *Introduction to Metric and Topological Spaces* (Oxford: Clarendon Press).

Volterra, V., 1959, *Theory of Functionals and of Integral and Integro-Differential Equations* (New York: Dover).

3

ROBUST NEURAL
NETWORKS

James T. Lo

3.1 INTRODUCTION

Risk-sensitive criteria, which were first proposed by Jacobson (1973)
and Speyer et al. (1974) for stochastic linear control, are known to
be closely related with deterministic games and H_∞ criteria for robust
linear control and estimation. In fact, such relations have attracted
a great deal of attention in the past few years (Glover and Doyle
1988; Hassibi et al. 1996; Speyer et al. 1992; Whittle 1990). The risk-
sensitive criteria emphasize greater errors in an exponential manner
and thereby induce robust performances.

In this chapter, we generalize the standard risk-sensitive functional (i.e., exponential quadratic functional) and observe that risk-

This work was supported in part by the Computational Intelligence and Knowledge
Modelling Program, National Science Foundation, under Grant No. ECS-9707206.
Part of this chapter is reprinted with permission from the *Proceedings of the 1998
International Joint Conference on Neural Networks*, pp. 1311–1314 and pp.
2429–2434, IEEE Press, 1998.

sensitive functionals, except the risk-neutral ones (i.e., L_p norms), do not qualify as norms for a space of functions. In fact, they do not satisfy either homogeneity or triangle inequality required of a norm. Nevertheless, capabilities of neural networks to approximate functions and dynamic systems with respect to risk-sensitive criteria under mild regularity conditions are established here. These universal approximation capabilities qualify neural networks as powerful vehicles for robust processing (e.g., signal processing and control).

It is stated and proven that under mild conditions, a function can be approximated to any accuracy with respect to a general risk-sensitive functional by a multilayer perceptron (MLP) with one hidden layer of neurons, whose activation functions are either continuous or nondecreasing sigmoidal functions.

The proof of this universal risk-sensitive approximation capabilty of multilayer perceptrons is mainly based on the known fact that any continuous function can be approximated uniformly to an arbitrary accuracy by an MLP of the same type (Cybenko 1989; Funahashi 1989; Hornik et al. 1989). It is also known that many widely used radial-basis functions (RBFs) such as the Gaussian RBFs can also approximate any continuous function to an arbitray accuracy (Lo 1972; Parzen 1962; Poggio and Girosi 1989). This fact enables us to prove a similar universal risk-sensitive approximation capabilty of such RBFs. The proof is almost verbatim the same as that for multilayer perceptrons and is thus omitted here.

Of course, universal risk-sensitive approximation capabilities of other neural networks such as the high-order or $\Sigma\Pi$ networks, that can uniformly approximate continous functions arbitrarily closely, can similarly be estabilished.

It was proven in Lo (1993) that under mild regularity conditions, a dynamic system can be approximated to any accuracy with respect to an L_p criterion by a recurrent neural network in both the series-parallel and parallel formulations. This universal approximation capability justifies the enthusiasm over the recent development on the use of neural networks for system identification (Fernandez et al. 1990; Narendra and Parthasarathy 1990, 1991; Puskorious and Feldkamp 1992).

Essentially, three neural network paradigms have been employed for system identification. The first is the multilayer perceptrons with tapped-delay lines (MLPWTDs). They are used as building blocks in both the series-parallel and the parallel formulations in Narendra and Parthasarathy (1990, 1991).

The second paradigm that has been employed for system identification (Fernandez et al. 1990; Lo 1993; Puskorious and Feldkamp 1992) is the multilayer perceptrons with the neurons in each hidden layer interconnected through a unit time-delay device. A special kind of such networks was first suggested in Elman (1988). We denote such networks by MLPWINs (multilayer perceptron with interconnected neurons). In Puskorious and Feldkamp (1992), an example nonlinear system was identified by a series-parallel formulation using an MLPWIN. It was noted in Puskorious and Feldkamp (1991) that the numerical test results for MLPWINs are significantly superior to those achieved by MLPWTDs.

In Lo (1993), MLPs with free and teacher-forced output feedbacks, to be denoted by MLPWOFs, were proposed for system identification. Both the MLPWTD and the MLPWOF use a pure MLP (without output feedbacks or interconnects among the hidden neurons) as the principal component. The only difference between them is that the MLPWOF has output feedbacks instead of tapped-delay lines for the inputs. After proper training, the free feedbacks of the MLPWOF carry the optimal information for minimizing its external output errors and are hence more efficient than the tapped-delay lines for storing and processing the information.

In all these existing results on system identification by neural networks, the identification error criteria are the standard mean-square errors of the neural identifiers' outputs. In Lo (1993), it is proven that both an MLPWIN and an MLPWOF exist that approximate almost any dynamic system to any degree of accuracy with respect to the mean-square errors in the series-parallel identification formulation and, for a finite time period and under some more regularity conditions on the system, in the parallel formulation as well.

In this chapter, it is proven that under mild regularity conditions, a dynamic system can be approximated to any accuracy with respect to a risk-sensitive error criterion by an MLPWOF and an MLPWIN in both the series-parallel formulation and the parallel formulation.

To simplify our presentation, only SISO (single input–single output) functions and SISO dynamic systems being approximated by neural networks with a single hidden layer of neurons are treated in this chapter. However, all the discussions and results can be easily generalized to MIMO (multiple input–multiple output) systems being approximated by neural networks with more than one hidden layer of neurons.

3.2 PRELIMINARIES

Definition 1. A function $a: R^1 \to R^1$ is said to be sigmoidal, if

$$a(x) \to \begin{cases} 1 & as\ x \to +\infty \\ -1 & as\ x \to -\infty \end{cases}$$

The following theorem on the uniform approximation of continuous functions by MLPs is a combination of two theorems proven in Cybenko (1989), Funahashi (1989), and Hornik et al. (1989).

Theorem 1. Let a be a continuous or nondecreasing sigmoidal function, and let K denote a compact set in R^n. Given any continuous function $f: R^n \to R^1$ and an arbitrary $\varepsilon > 0$, there are an integer N, real constants c_i, w_{ij}, b_i, $i = 1, \ldots, N$, $j = 1, \ldots, n$, such that

$$\left| f(x_1, \ldots, x_n) - \sum_{i=1}^{N} c_i a\left(\sum_{j=1}^{n} w_{ij} x_j + b_j \right) \right| < \varepsilon$$

for all $(x_1, \ldots, x_n) \in K$.

The following theorem on the uniform approximation of continuous functions by RBFs is an immediate consequence of the classic Stone–Weierstrass theorem or Wiener approximation theorem (Lo 1972; Parzen 1962; Poggio and F. Girosi 1989).

Theorem 2. Let $a(r, w_0)$ be the Gaussian function $\exp(-r^2/w_0^2)$, the thin-plate-spline function $r^2 \ln r$, the multiquadric function $(r^2 + w_0^2)^{1/2}$, or the inverse multiquadric function $(r^2 + w_0^2)^{-1/2}$, where w_0 is a parameter and r is a real variable. Let K denote a compact set in R^n. Given any continuous function $f: R^n \to R^1$ and an arbitrary $\varepsilon > 0$, there are an integer N, real constants w_0, c_i, w_{ij}, $i = 1, \ldots, N$, $j = 1, \ldots, n$, such that

$$\left| f(x_1, \ldots, x_n) - \sum_{i=1}^{N} c_i a\left(\sqrt{\sum_{j=1}^{n} (x_j - w_{ij})^2}, w_0 \right) \right| < \varepsilon$$

for all $(x_1, \ldots, x_n) \in K$.

3.3 GENERAL RISK-SENSITIVE FUNCTIONALS

Definition 2. Let p be a positive number and λ be a real number, and $S_p(K)$ be the set of real-valued functions f defined on a compact set K in R^n with probability measure $\mu(K) = 1$ such that the Lebesgue integral,

$$E\left[\exp\left(\lambda|f|^p\right)\right] := \int \exp\left(\lambda|f(x)|^p\right)\mu(dx) < \infty$$

The order-(λ,p) risk-sensitive functional $\rho_{\lambda,p}$ on $S_p(K)$ is defined to be

$$\rho_{\lambda,p}(f) := \frac{1}{\lambda}\ln\int\exp\left(\lambda|f(x)|^p\right)\mu(dx)$$

It is shown that as $\lambda \to 0$, $\rho_{\lambda,p}(f)$ approaches $E(|f|^p)$. We define $\rho_{0,p}(f)$ to be the functional $E(|f|^p)$.

The order-$(\lambda,2)$ risk-sensitive functional for $\lambda > 0$ was proposed by Jacobson (1973) and is called the standard risk-sensitive functional. We note that by the convexity of $(\cdot)^a$ for $a > 1$ [or the concavity of $(\cdot)^a$ for $a < 0$], Jensen's inequality, and the monotonicity of $\ln(\cdot)$, for $\lambda > 0$ and $c > 0$ (or for $\lambda < 0$ and $0 < c < 1$),

$$\frac{1}{\lambda}\ln E\left[\exp \lambda|cf|^p\right] = \frac{1}{\lambda}\ln E\left[\left(\exp \lambda|f|^p\right)^{c^p}\right]$$

$$\geq \frac{1}{\lambda}\ln\left(E\left[\exp \lambda|f|^p\right]\right)^{c^p}$$

$$= \frac{c^p}{\lambda}\ln\left(E \exp \lambda|f|^p\right)$$

whence for $\lambda > 0$ and $c > 1$ (or for $\lambda < 0$ and $0 < c < 1$),

$$\rho_{\lambda,p}(cf) \geq c^p\rho_{\lambda,p}(f)$$
$$[\rho_{\lambda,p}(cf)]^{1/p} \geq c[\rho_{\lambda,p}(f)]^{1/p}$$

where the qualities hold if and only if f is a constant with probability one. This shows that neither $\rho_{\lambda,p}$ nor $\rho_{\lambda,p}^{1/p}$ satisfies homogeneity for $\lambda > 0$ and $c > 1$ (or for $\lambda < 0$ and $0 < c < 1$), and that neither $\rho_{\lambda,p}$ nor

$\rho_{\lambda,p}^{1/p}$ satisfies triangle inequality for $\lambda > 0$ and $c > 1$. Note that homogeneity and triangle inequality are properties required of a norm.

Theorem 3. If $E[\exp \lambda |f|^p] < \infty$ for some p and λ, then $E[\exp \alpha |f|^q] < \infty$ for any α and $q < p$.

Proof. Let $S := \{x: |f(x)|^{p-q} \geq \alpha/\lambda\}$ and note that

$$E[\exp \alpha |f|^q] = \left(\int_{S+} \int_{R^n \setminus S}\right) \exp \alpha |f(x)|^q \mu(dx)$$

$$\leq \int_S \exp \lambda |f(x)|^p \mu(dx)$$

$$+ \int_{R^n \setminus S} \exp \alpha \left|\frac{\alpha}{\lambda}\right|^{q/(p-q)} \mu(dx)$$

$$\leq \infty$$

Definition 3. Given a function $f: R^n \to R^1$, a truncated version f_M is defined by

$$f_M(x) = \begin{cases} f(x) & \text{if } |f(x)| \leq M \\ M & \text{if } f(x) > M \\ -M & \text{if } f(x) < -M \end{cases}$$

for a positive $M > 0$.

Note that $|f(x) - f_M(x)| \leq |f(x)|$ for all $x \in R^n$.

3.4 APPROXIMATION OF FUNCTIONS BY MLPs

In this section, we state and prove the main thorem.

Theorem 4. Let a be a continuous or nondecreasing sigmoidal function, and let K, μ, $\rho_{\lambda,p}$, and E be defined as in Definition 2. Let $f: R^n \to R^1$ be a Borel measurable function. If $\lambda > 0$, assume that $\rho_{\lambda c_p r, p}(f) < \infty$, where r is some number greater than 1, and $c_p = 1$ for $0 < p \leq 1$, and $= 2^{p-1}$ for $p > 1$. Given any $\varepsilon > 0$, there are an integer N, real constants c_i, b_i, $i = 1, \ldots, N$, and row n-vectors $w_i = [w_{i1}, \ldots, w_{in}]$, $i = 1, \ldots, N$, such that

$$\rho_{\lambda,p}(f - s) < \varepsilon \tag{3.1}$$

$$s(x) := \sum_{i=1}^{N} c_i a(w_i.x + b_i) \qquad (3.2)$$

Proof. Case 1: $\lambda > 0$: By Theorem 1, there is a sequence

$$s_k(x) := \sum_{i=1}^{N} c_i(k) \, a \, (w_i(k)x + b_i(k)) \qquad (3.3)$$

such that $\lim_{k \to \infty} s_k(x) = f_M(x)$ and $|s_k(x)| < M$ for almost every $x \in K$ and for any positive M. Notice that $|f_M(x) - s_k(x)| \le 2M$ and hence $E \exp[\lambda c_p r' |f_M - s_k|^p] \le \exp[2^p \lambda c_p r' M^p]$ for $r' = r/(r-1)$. By the dominated convergence theorem,

$$\lim_{k \to \infty} \rho_{\lambda c_p r', p}(f_M - s_k) = 0 \qquad (3.4)$$

Now notice that $|f(x) - f_M(x)|^p \le |f(x)|^p$ and hence $\rho_{\lambda c_p r, p}(f - f_M) \le \rho_{\lambda c_p r, p}(f)$. Since $\lim_{M \to \infty} |f(x) - f_M(x)|$ for all $x \in R^n$, by the dominated convergence thorem, the assumption that $\rho_{\lambda c_p r, p}(f) < \infty$ implies that

$$\lim_{M \to \infty} \rho_{\lambda c_p r, p}(f - f_M) = 0 \qquad (3.5)$$

We are now ready to show (3.1) in the following: By an algebraic inequality and the Hölder inequality, we obtain for $\lambda > 0$,

$$\begin{aligned}
\rho_{\lambda, p}(f - s_k) &= \rho_{\lambda, p}(f - f_M + f_M - s_k) \\
&\le \frac{1}{\lambda} \ln\Big[E \exp\big(\lambda c_p |f - f_M|^p \\
&\quad + \lambda c_p |f_M - s_k|^p \big) \Big] \\
&\le \frac{1}{\lambda} \ln\big(E\big[\exp \lambda c_p r |f - f_M|^p \big]\big)^{1/r} \\
&\quad \big(E\big[\exp \lambda c_p r' |f_M - s_k|^p \big]\big)^{1/r'} \\
&= \frac{1}{\lambda r} \ln E\big[\exp \lambda c_p r |f - f_M|^p \big] \\
&\quad + \frac{1}{\lambda r'} \ln E\big[\exp \lambda c_p r' |f_M - s_k|^p \big]
\end{aligned}$$

It follows from (3.4) and (3.5) that for any $\varepsilon > 0$, there are M and k such that $\rho_{\lambda, p}(f - s_k) < \varepsilon$. This completes the proof for $\lambda > 0$.

Case 2: $\lambda < 0$. By Theorem 1, there is a sequence

$$s_k(x) = \sum_{i=1}^{N} c_i(k)\, a\, (w_{i\cdot}(k)x + b_i(k)) \tag{3.6}$$

such that $\lim_{k\to\infty} s_k(x) = f(x)$ for almost all $x \in K$. Note that $0 < \exp \lambda |f - s_k|^p < 1$. By the dominated convergence theorem, $\lim_{k\to\infty}\rho_{\lambda,p}(f - s_k) = 0$, completing the proof for $\lambda < 0$.

3.5 APPROXIMATION OF FUNCTIONS BY RBFs

We notice that the proof of the main theorem in the preceding section is based on Theorem 1 concerning the uniform approximation by MLPs. Theorem 2 is analogous to Theorem 1 and thus provides a basis for extension of Theorem 4. Since the proof of the extension is almost verbatim the same, we only state the extended theorem for risk-sensitive approximation of functions by RBFs in the following.

Theorem 5. Let $a(r, w_0)$ be defined as in Theorem 2, and let $K, \mu, \rho_{\lambda,p}$, and E be defined as in Definition 2. Let $f\colon R^n \to R^1$ be a Borel measurable function. If $\lambda > 0$, assume that $\rho_{\lambda c_p r, p}(f) < \infty$, where r is some number greater than 1, and $c_p = 1$ for $0 < p \le 1$, and $= 2^{p-1}$ for $p > 1$. Given any $\varepsilon > 0$, there are an integer N, real constants w_0, c_i, w_{ij}, $i = 1, \ldots, N, j = 1, \ldots, n$, such that

$$\rho_{\lambda,p}(f - s) < \varepsilon \tag{3.7}$$

$$s(x) := \sum_{i=1}^{N} c_i a\left(\sqrt{\sum_{j=1}^{n}(x_j - w_{ij})^2}, w_0\right) \tag{3.8}$$

3.6 FORMULATION OF RISK-SENSITIVE IDENTIFICATION OF SYSTEMS

Consider the discrete-time SISO causal time-invariant plant

$$y(t+1) = f(y(t), y(t-1), \ldots, y(t-p+1), \tag{3.9}$$

$$u(t), u(t-1), \ldots, u(t-q+1)) \tag{3.10}$$

where $(u(t), y(t))$ is the input–output pair at time t. The finite integers p and q are unknown. It is assumed that for all t, $|u(t)| < B$, $|y(t)| < B$ and $f(B^{p+q}) \subset (-B, B)$ for some known $B < \infty$ and that $p < b$ and $q < b$ for some known $b < \infty$. No mathematical description for the function f is available. Nevertheless, it is well defined for the specification of the plant by all the possible sequences of the input and output pairs $(u(t), y(t))$.

The problem of identifying the plant (3.10) is to construct a mathematical system, called a system identifier, whose output $\hat{y}(t + 1)$ at time t approximates that of the plant as closely as possible during the operations of the plant, using either the operational data or the experimental data or both. The data, whether operational or experimental, must be sufficiently representative of all the possible operational sequences of input–output pairs and sufficiently reflective of the joint probability distributions of u and y viewed as stochastic processes. We assume that both u and y are stationary stochastic processes.

For risk-sensitive identication of the plant (3.10), the order-(λ, p) risk-sensitive criterion is

$$\frac{1}{\lambda} \ln E\left[\exp \lambda \sum_{t=b+1}^{T} |y(t) - \hat{y}(t)|^p\right] \tag{3.11}$$

where T is a positive integer selected such that the operations of the plant in the time interval $[b + 1, T]$ represent those of the plant over its typical continuous runs and $\hat{y}(t + 1)$ denotes the output of the system identifier at time t. This criterion is similar to the control and estimation criteria used in Jacobson (1973) and Speyer et al. (1992). An alternative order-(λ, p) risk-sensitive criterion is

$$\frac{1}{\lambda} \ln E\left[\frac{1}{T - b} \sum_{t=b+1}^{T} \exp \lambda |y(t) - \hat{y}(t)|^p\right] \tag{3.12}$$

The theorems to be stated and proven in the sequel are valid with the criterion (3.11) replaced with the criterion (3.12). It is easy to see how the given proofs are modified to suit (3.12).

There are essentially two formulations for dynamic system identification, namely the series-parallel formulation and the parallel formulation. In the series-parallel formulation, the system identifier inputs $u(t)$ and $y(t)$ at time t. In the parallel formulation, the system identifier inputs $u(t)$ and $\hat{y}(t)$ at time t.

3.7   SERIES-PARALLEL IDENTIFICATION BY ARTIFICIAL NEURAL NETWORKS (ANNs)

Throughout the rest of this chapter, the activation function a of every neuron is assumed to be a strictly increasing sigmoidal function.

Theorem 6. Consider the series-parallel identification of the plant (3.10). Given $\varepsilon > 0$, there are an MLPWTD, MLPWOF, and MLPWIN that input $u(t-1)$ and $y(t-1)$ and outputs $\alpha(t)$ at time t such that

$$\frac{1}{\lambda} \ln E\left[\exp \lambda \sum_{t=b+1}^{T} |y(t) - \alpha(t)|^p\right] < \varepsilon \qquad (3.13)$$

for any given real number λ and positive real number p.

Proof. We will first prove the existence of an MLPWTD such that (3.13) holds. The proof will then be extended to the existence of an MLPWOF and MLPWIN.

Case 1: $\lambda > 0$. Consider the function $f_0(\xi)\colon = f(\eta)$, where $\xi\colon = [\xi_1, \ldots, \xi_{2b}]$ and $\eta\colon = [\xi_1, \ldots, \xi_r, \xi_{b+1}, \ldots, \xi_{b+q}]^T$. Note that by this definition of f_0 and the boundedness assumption on f, we have $|f_0(B^{2b})| = |f(B^{r+q})| \subset (-B, B)$. By Theorem 4, given $\varepsilon > 0$, an MLP, which is a function $g\colon B^{r+q} \to R$, can be selected such that the error $\rho_{\lambda(T-b),p}(f_0 - g)$, where the probability measure μ is defined by $y(t), \ldots, y(t-r+1)$, $u(t), \ldots, u(t-q+1)$, is less than $\varepsilon/(T-b)$ at each time $t+1$.

Let two tapped-delay lines hold $u(t), \ldots, u(t-b+1)$ and $y(t), \ldots, y(t-b+1)$ at time $t+1$ and let the selected MLP be used to transform them into $\alpha(t+1) = g(y(t), \ldots, y(t-b+1), u(t), \ldots, u(t-b+1))$. By a generalized Hölder inequality, the risk-sensitive identification criterion (3.11) satisfies

$$\frac{1}{\lambda} \ln E \exp \lambda \sum_{t=b+1}^{T} |y(t) - \alpha(t)|^p$$

$$\leq \frac{1}{\lambda} \ln \prod_{t=b+1}^{T} \left[E \exp \lambda(T-b) |y(t) - \alpha(t)|^p\right]^{1/(T-b)}$$

$$= \frac{1}{\lambda(T-b)} \sum_{t=b+1}^{T} \ln E \exp \lambda(T-b) |y(t) - \alpha(t)|^p$$

$$= (T-b)\, \rho_{\lambda(T-b),p}(f_0 - g) < \varepsilon$$

Case 2: $\lambda \leq 0$. Let f_0 be defined as before. Given $\varepsilon > 0$, an MLP, which is a function $g: B^{r+q} \to R$, can be selected such that the error $E|f_0 - g|^p$, where the probability measure μ is defined by $y(t), \ldots,$ $y(t - r + 1), u(t), \ldots, u(t - q + 1)$, is less than $\varepsilon/(T - b)$ at each time $t + 1$. Then the risk-sensitive identification criterion (3.11) satisfies

$$\frac{1}{\lambda} \ln E \exp \lambda \sum_{t=b+1}^{T} |y(t) - \alpha(t)|^p$$

$$\leq \frac{1}{\lambda} \ln \exp \lambda \sum_{t=b+1}^{T} E |y(t) - \alpha(t)|^p$$

$$= \sum_{t=b+1}^{T} E |f_0 - g|^p < \varepsilon$$

This completes the proof for $\lambda \leq 0$.

We will now extend the above proof to the existence of an MLPWOF and MLPWIN. Connect $2b - 2$ hidden neurons in series in such a way that these neurons and the input terminals of the ANN (MLPWOF or MLPWIN) form two tapped-delay lines for holding $u(t), \ldots, u(t - b + 1)$ and $y(t), \ldots, y(t - b + 1)$ at time $t + 1$. The rest of the proof is identical to the proof above for the existence of an MLPWTD.

3.8 PARALLEL IDENTIFICATION BY ANNs

Theorem 7. Consider the plant (3.10) with the range of f contained in $(-B, B)$ for some $B < \infty$ and with the $L_{p'}$ norm $\|\nabla_y f(y(t), \ldots, y(t - r + 1), u(t), \ldots, u(t - q + 1))\|_{p'}$ of the gradient $\nabla_y f$ with respect to $(y(t), \ldots, y(t - r + 1))$ uniformly bounded by M, where $p' = p/(p - 1)$. Assume that for all t, $|u(t)| < B$ and $|y(t)| < B$ and that the finite integers r and q are less than or equal to $b < \infty$. Given $\varepsilon > 0$, there exist an appropriately initialized MLPWTD, MLPWOF and MLPWIN, all with a single hidden layer of neurons, such that

$$\frac{1}{\lambda} \ln E \left[\exp \lambda \sum_{t=b+1}^{T} |y(t) - \alpha(t)|^p \right] < \varepsilon \qquad (3.14)$$

where $\alpha(t)$ denotes the output of the MLPWOF at time t, which has taken $\{(u(\tau), \alpha(\tau)), \tau = b + 1, \ldots, t - 1\}$ as the inputs in the given order.

Proof. We will first prove the existence of an MLPWOF. How the proof is extended for the existence of an MLPWIN will then be discussed. The extension of the proof to an MLPWTD is trivial and thus omitted.

Let an MLPWOF with $J := r + q - 2$ free output feedbacks, $\beta_1(t)$, $\ldots, \beta_J(t)$ be initialized as follows: At time $t = 0$, set $\beta_1(t + 1) = \cdots = \beta_J(t + 1) = B$ and $\alpha(t + 1) = y(1)$ at the output terminals for feedbacks. At times $t = 1, \ldots, b - 1$, set $\alpha(t + 1) = y(t + 1)$ at the teacher-forced output terminal and $u(t) = u(t)$ at the external input terminal. Here the first $u(t)$ denotes the input terminal and the second $u(t)$ denotes the input at time t. At this point, the MLPWOF is ready to run by itself, starting with taking the input $u(b)$ and producing the output $\alpha(b + 1)$.

Construct $r - 1$ free feedback loops so that for $t = b, \ldots, T - 1$ and $\tau = 2, \ldots, r - 1$,

$$\beta_\tau(t) = a(\beta_{\tau-1}(t - 1)) \tag{3.15}$$

and

$$\beta_1(t) = a(\alpha(t)) \tag{3.16}$$

Construct $q - 1$ additional free feedback loops so that for $t = b, \ldots, T - 1$ and $\tau = r + 1, \ldots, r + q - 2$,

$$\beta_\tau(t) = a(\beta_{\tau-1}(t - 1)) \tag{3.17}$$

and

$$\beta_r(t) = a(u(t)) \tag{3.18}$$

At time $t + 1$, the input vector to the MLPWOF is

$$z(t) := [\alpha(t), a(\alpha(t-1)), a^{\circ 2}((\alpha(t-2)), \ldots, \tag{3.19}$$

$$a^{\circ(r-1)}(\alpha(t-r+1)), u(t), a(u(t-1)), \tag{3.20}$$

$$a^{\circ 2}(u(t-2)), \ldots, a^{\circ(q-1)}(u(t-q+1))]^T \tag{3.21}$$

where $a^{\circ \tau}$ denotes the composition $a \circ a \circ \cdots \circ a$ of τ copies of a. Notice that $z(t)$ can be viewed as a random vector whose range is contained in the compact set $K := [-B,B] \times [-1,1]^{r-1} \times [-B,B] \times [-1,1]^{q-1}$.

The input terminals receiving the input vector (3.21), all the hidden neurons except those used above to construct the free feedback loops, and the eternal output terminals form an MLP inside the MLPWOF. Denoting the function representing the MLP by g, the external output of the MLPWOF at time $t+1$ is

$$\alpha(t+1) = g(\alpha(t), a(\alpha(t-1)), \ldots, a^{\circ(r-1)}(\alpha(t-r+1)),$$
$$u(t), a(u(t-1)), \ldots, a^{\circ(q-1)}(u(t-q+1)))$$

In the following, we shall first establish an inequality (3.22). The following symbols will be needed:

$$u^t := (u(t), u(t-1), \ldots, u(t-q+1))$$
$$\alpha^t := (\alpha(t), \alpha(t-1), \ldots, \alpha(t-r+1))$$
$$y^t := (y(t), y(t-1), \ldots, y(t-r+1))$$
$$f(y^t, u^t) := f(y(t), \ldots, y(t-r+1), u(t), \ldots, u(t-q+1))$$
$$p(\alpha^t, u^t) := g(\alpha(t), a(\alpha(t-1)), \ldots, a^{\circ(r-1)}(\alpha(t-r+1)), u(t),$$
$$a(u(t-1)), \ldots, a^{\circ(q-1)}(u(t-q+1)))$$
$$c_p := 1, \text{ for } 0 < p \le 1 \text{ and} := 2^{p-1} \quad \text{for } 1 < p$$
$$d_t := \alpha(t) - y(t)$$
$$\delta^t := p(\alpha^t, u^t) - f(\alpha^t, u^t)$$
$$a_i := c_p (c_p M^p)^{i-1} \sum_{t=b-i+1}^{T-i} \sum_{\tau_1 = t-b+1}^{t} \cdots \sum_{\tau_{i-1}=\tau_{i-2}-b+1}^{\tau_{i-2}} |\delta^{\tau_{i-1}}|^p$$

Note that $y(t+1) = f(u^t, y^t)$ and $\alpha(t+1) = p(u^t, \alpha^t)$.

By algebraic inequalities, the Hölder inequality, a mean value theorem and the given $L_{p'}$ bound M of the gradient $\nabla_y f$, we have

$$|d_{t+1}|^p \le [|f(y^t, \alpha^t) - f(y^t, u^t)| + |\delta^t|]^p$$
$$\le c_p |f(y^t, \alpha^t) - f(y^t, u^t)|^p + a_1$$
$$\le c_p |\nabla_y f(y^t + v(\alpha^t - y^t), u^t)(\alpha^t - y^t)|^p + c_p |\delta^t|^p$$
$$\le c_p \|\nabla_y f(y^t + v(\alpha^t - y^t), u^t)\|_{p'}^p \|\alpha^t - y^t\|_p^p + c_p |\delta^t|^p$$
$$\le c_p M^p \sum_{\tau = t-b+1}^{t} |d_\tau|^p + c_p |\delta^t|^p$$

Applying this inequality repeatedly and noting that $d_t = 0$ for $t \le b$ yield

$$\sum_{t=b+1}^{T} |d_t|^p \le c_p M^p \sum_{t=b}^{T-1} \sum_{\tau_1=t-b+1}^{t} |d_{\tau_1}|^p + a_1$$

$$\le (c_p M^p)^2 \sum_{t=b-1}^{T-2} \sum_{\tau_1=t-b+1}^{t} \sum_{\tau_2=\tau_1-b+1}^{\tau_1} |d_{\tau_2}|^p$$

$$+ \sum_{i=1}^{2} a_i \dots$$

$$\le (c_p M^p)^{T-b-1} \sum_{t=b-(T-b)}^{T-(T-b-1)} \sum_{\tau_1=t-b+1}^{t} \dots$$

$$\sum_{\tau_{T-b-1}=\tau_{T-b-2}-b+1}^{\tau_{T-b-2}} |d_{\tau_{T-b-1}}|^p + \sum_{i=1}^{T-b-1} a_i$$

$$= (c_p M^p)^{T-b-1} |d_{b+1}|^p + \sum_{i=1}^{T-b-1} a_i$$

$$\le c_p (c_p M^p)^{T-b-1} |\delta^b|^p + \sum_{i=1}^{T-b-1} a_i$$

Notice that

$$\sum_{i=1}^{T-b-1} a_i = c_p \sum_{i=1}^{T-b-1} \sum_{t=b+1}^{T-i} (c_p M^p b)^{i-1} |\delta^t|^p$$

$$= c_p \sum_{t=b+1}^{T-1} \left[\sum_{i=1}^{T-t} (c_p M^p b)^{i-1} \right] |\delta^t|^p$$

$$= c_p \sum_{t=b+1}^{T-1} \left[(c_p M^p b)^{T-t} - 1 \right]$$

$$|\delta^t|^p / (c_p M^p b - 1)$$

Denoting the greater of $[c_p(c_p M^p b)^{T-t} - 1]/(c_p M^p b - 1)$ and $c_p(c_p M^p)^{T-b-1}$ by K, we obtain the desired inequality

$$\sum_{t=b+1}^{T} |d_t|^p \le K \sum_{t=b}^{T-1} |\delta^t|^p \qquad (3.22)$$

Case 1: $\lambda > 0$. By a generalized Hölder inequality, it follows from (3.22) that

$$\frac{1}{\lambda} \ln E \exp \lambda \sum_{t=b+1}^{T} |d_t|^p \tag{3.23}$$

$$\leq \frac{1}{\lambda(T-b)} \sum_{t=b}^{T-1} \ln E \exp(T-1)\lambda K |\delta^t|^p$$

$$= \frac{1}{\lambda} \ln E \exp(T-1)\lambda K |\delta^t|^p \tag{3.24}$$

By Theorem 4, given any $\varepsilon > 0$, there is an MLP represented by the function g such that

$$\frac{1}{\lambda} \ln E \exp(T-1)\lambda K |\delta^t|^p < \varepsilon \tag{3.25}$$

Combining (3.24) and (3.25) completes the proof for $\lambda > 0$.

Case 2: $\lambda \leq 0$. By the well-known L_p function approximation capability of the MLPs, given any $\varepsilon > 0$, there is an MLP represented by the function g such that $E|\delta|^p < \varepsilon/(T-b)K$. By (3.22) and Jensen's inequality,

$$\frac{1}{\lambda} \ln E \exp \lambda \sum_{t=b+1}^{T} |d_t|^p \leq \frac{1}{\lambda} \ln E \exp \lambda K \sum_{t=b}^{T-1} |\delta^t|^p$$

$$\leq \frac{1}{\lambda} \ln \exp \lambda K \sum_{t=b}^{T-1} |\delta^t|^p$$

$$\leq \varepsilon$$

completing the proof for $\lambda \leq 0$.

We will now consider the proof of the existence of an MLPWIN such that (3.14) holds. An MLPWIN has only one input terminal reserved for the exogenous input $u(t)$. Then how do we initialize the MLPWIN to incorporate $u(b-1), \ldots, u(1), y(b), \ldots, y(1)$ and then feed back the external output $\alpha(t)$ into the network? One way is to designate a hidden neuron, say neuron 1, for these tasks, which we call the initialization neuron. The same weights for the output terminal $\alpha(t+1)$ are assigned to the interconnects leading into the initialization neuron. Thus the activation level $\beta_1(t + 1)$ of the initialization neuron is equal to $a(\alpha(t + 1))$. After a unit time delay, it [i.e., $a(\alpha(t))$] is distributed as if it were an input terminal. Since a is strictly monotone increasing, the feedback $a(\alpha(t))$ carries as much information as $\alpha(t)$.

We use $r - 1$ hidden neurons to construct a feedback loop for holding $a^{o(i+1)}(\alpha(t - i))$ for $i = 1, \ldots, r - 1$, and $q - 1$ hidden neurons to construct a feedback loop for holding $a^{o(i+1)}(u(t - i))$ for $i = 1, \ldots, q - 1$. These feedback loops are similar to the two free feedback loops constructed within the foregoing MLPWOF.

The MLPWIN is initialized as follows: At time $t = 0$, set $\beta_1(t + 1) = a(y(t + 1))$ and set the activation levels of all the other neurons in the MLPWIN, $\beta_j(t + 1) = B$ for $j = 2, \ldots, r$, and $t = 1, \ldots, r$. At times $t = 1, \ldots, b - 1$, set $\beta_1(t + 1) = a(y(t + 1))$ at the initialization neuron and $u(t) = u(t)$ at the external (the only) input terminal. At time $t = b$, the MLPWIN is ready to run by itself by starting with taking the first input $u(b)$ and producing the first output $\alpha(b + 1)$.

From this point onward, the proof for the existence of an MLPWIN such that (3.14) holds is almost verbatim the same as the above proof for the existence of an MLPWOF.

3.9   CONCLUSION

This chapter generalizes the standard risk-sensitive (exponential quadratic) criteria and shows that under mild conditions, functions can be approximated to any accuracy with respect to general risk-sensitive functionals by MLPs and RBFs. These results are mainly based on these neural networks' capabilities to uniformly approximate continuous functions arbitrarily closely on a compact support.

These risk-sensitive approximation results can therefore be extended to other neural network paradigms that have similar uniform approximation capabilities. Among such neural networks paradigms are variations of the MLPs and RFBFs considered here and the high-order neural networks (Hassoun 1995; Haykin 1998).

Based on these risk-sensitive approximation results, this chapter shows that under mild conditions, a dynamic system can be approximated or identified in both the series-parallel and parallel formulation, to any accuracy with respect to general risk-sensitive criteria, by an MLPWTD, an MLPWOF, and/or an MLPWIN. Obviously, these risk-sensitive identification results can be extended to other neural network paradigms in the same way that the foregoing risk-sensitive approximation results are extended.

In view of the difficulties in designing robust nonlinear systems for control, communication, and signal processing by the traditional analytic approach as opposed to the synthetic approach that neural networks represent, the universal risk-sensitive approximation and identification capabilities of the neural networks offer a most promising approach to robust processing in control, communication, and signal processing.

REFERENCES

Cybenko, G., 1989, "Approximation by superpositions of a sigmoidal function," *Math. Control, Signals Syst.*, vol. 2, pp. 303–314.

Elman, J. L., 1988, "Finding structure in time," Technical Report CRL 8801, Center for Research in Language, University of California, San Diego.

Fernandez, B., A.G. Parlos, and W. K. Tsai, 1990, "Nonlinear dynamic system identification using artifitial neural networks (ANNs)," *Proc. IJCNN-90*, San Diego, CA, pp. 133–141.

Funahashi, K., 1989, "On the approximate realization of continuous mappings by neural networks," *Neural Networks*, vol. 2, pp. 183–192.

Glover, K., and J. C. Doyle, 1988, "State-space formulae for all stabilizing, controllers that satisfy an h-infinity norm bound and relations to risk-sensitivity," *Syst. Control Lett.*, vol. 11, pp. 167–172.

Hassibi, B., A. H. Sayed, and T. Kailath, 1996, "H-infinity optimality of the LMS algorithm," *IEEE Trans. Signal Processing*, vol. 44, pp. 267–280.

Hassoun, M. H., 1995, *Fundamentals of Artificial Neural Networks* (Cambridge, MA: MIT Press).

Haykin, S., 1998, *Neural Networks—A Comprehensive Foundation*, Second Edition (New Jersey: Prentice-Hall).

Hornik, K., M. Stinchcombe, and H. White, 1989, "Multilayer feedforward networks are universal approximators," *Neural Networks*, vol. 2, pp. 259–366.

Jacobson, D. H., 1973, "Optimal stochastic linear systems with exponential performance criteria and their relation to deterministic games," *IEEE Trans. Automatic Control*, vol. AC-18-2, pp. 124–131.

Lo, J. T., 1972, "Finite dimensional sensor orbits and optimal nonlinear filtering," *IEEE Trans. Information Theory*, vol. IT-18-15, pp. 583–588.

Lo, J. T., 1993, "Dynamical system identification by recurrent multilayer perceptron," *Proc. of the 1993 World Congress on Neural Networks*, Portland, Oregon.

Narendra, K. S., and K. Parthasarathy, 1990, "Identification and control of dynamical systems using neural networks," *IEEE Trans. Neural Networks*, vol. 1, no. 1, pp. 4–27.

Narendra, K. S., and K. Parthasarathy, 1991, "Gradient method for the optimization of dynamical systems containing neural networks," *IEEE Trans. Neural Networks*, vol. 2, no. 2, pp. 252–262.

Parzen, E., 1962, "On estimation of a probability density function and mode," *Ann. Math. Stat.*, vol. 33, pp. 1065–1076.

Poggio, T., and F. Girosi, 1989, "A theory of networks for approximation and learning," Technical Report AI Memo No. 1140, Department of Mathematics, MIT, Cambridge, MA.

Puskorious, G. V., and L. A. Feldkamp, 1991, "Decoupled extended Kalman filter training of feedforward layered networks," *Proc. IJCNN-91*, Seatle, WA, pp. 771–777.

Puskorious, G. V., and L. A. Feldkamp, 1992, "Model reference adaptive control with recurrent networks trained by the dynamic DEKF algorithm," *Proc. IJCNN-92*, Baltimore, MD, pp. 106–113.

Speyer, J., J. Deyst, and D. H. Jacobson, 1974, "Optimaization of stochastic linear systems with additive measurement and process noise using exponential performance criteria," *IEEE Trans. Automatic Control*, vol. AC-19, pp. 358–366.

Speyer, J. L., C.-H. Fan, and R. N. Banavar, 1992, "Optimal stochastic estimation with exponential cost criteria," *Proc. 31st Conference on Decision and Control*, New York, New York.

Whittle, P., 1990, *Risk Sensitive Optimal Control* (New York: Wiley.)

4

MODELING, SEGMENTATION, AND CLASSIFICATION OF NONLINEAR NONSTATIONARY TIME SERIES

Craig L. Fancourt and Jose C. Principe

4.1 INTRODUCTION

In recent years there has been a growing interest in fields as diverse as statistics, engineering, biology, geology, and economics to soften the requirement of stationarity required by the traditional stochastic modeling of time series (Box and Jenkins 1976). It seems that some systems, rather than exhibiting a slow evolution, may exist in stable regimes for long periods of time and then, suddenly, exhibit a rapid, abrupt change to another regime. The goal of this chapter is to examine a new class of algorithms for detecting and characterizing abrupt changes in time series.

According to Klein's (1997) work on the history of time-series analysis, Darwin and Wallace's natural selection theory (1858) spawned some of the first mathematical investigations of nonstationarity, long before Khinchin (1932) and Wold (1938) defined it in modern statistical terms. Weldon (1890) observed that the distribution of the breadth of the forehead of Naples crabs formed a double peak and he, as well as Wallace, Pearson, and Galton, hypothesized that the presence of such bimodal distributions might be evidence of a bifurcation occurring over time. Although these measurements were made at a single moment in time, their problem bears a resemblance to the central problem considered in this chapter. Given a series of measurements over time that, in the aggregate, are observed to be multimodal, when does multimodality imply nonstationarity, or more precisely, piecewise stationarity?

As a simple example, consider a time series that switches between two memoryless Gaussian random processes with unit variance but with means of −1.5 and 1.5, respectively. Figure 4.1a shows a time series produced by random switching with equal probability at every

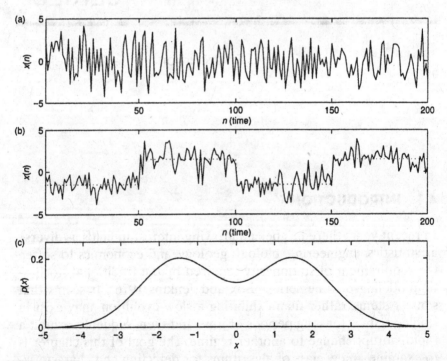

Figure 4.1 Multimodal time series: (a) random regime switching, (b) regime switching every 50 samples, and (c) pdf representing either time series.

time step; this process is strictly stationary. Figure 4.1*b* shows a time series where the processes switch every 50 samples; this process is nonstationary but locally piecewise stationary. And yet, both of these time series have identical multimodal distributions, as shown in Fig. 4.1*c*, and thus can be fit to the exact same Gaussian mixture model using a standard method such as the expectation–maximization algorithm (Dempster et al. 1977). However, if the mixture model is used with a properly tuned segmentation algorithm, the first time series can be recognized as stationary but multimodal while the second recognized as piecewise stationary. As we will see, the case of switching among processes with memory must be analyzed differently, but the approach is fundamentally the same.

The two fundamental issues in modeling and segmentation of time series are identification of a stationary segment and detection of a statistically significant change. We focus on a particular class of algorithms called mixture models that can model and segment piecewise stationary time series in a completely unsupervised fashion. They use an adaptive mixture of adaptive models that are local in signal space. The models have short-term memory sufficient for local modeling but no long-term memory sufficient to recognize piecewise stationarity. That is, given a piecewise stationary time series, these mixture models will learn equally well irrespective of any random permutation of the stationary segments. This invariance allows the modeling to be done independently of the segmentation. Once trained, the mixture models are analyzed with supervised segmentation algorithms to detect change points and identify the models within the time series. The segmentation information can in turn be used to refine the models.

We need to give more precise definitions of some of the terms used, but first it is instructive to consider some real data that can be regarded as at least piecewise quasi-stationary. At a small enough time scale, speech can be regarded as a sequence of quasi-stationary segments. These segments can be considered quasi-stationary because, although slowly evolving due to co-articulation effects, they have been successfully modeled by simple linear autoregressive (AR) systems with constant parameters, driven either by a pulse train, in the case of voiced speech, or noise, in the case of unvoiced speech.

There are many applications of segmentation and classification of signals. In the case of speech recognition, detection of the quasi-stationary segments results in improved estimates of associated feature vectors that in turn can reduce the required complexity of higher level word or subword recognizers. In monitoring any process, such as the operation of a jet or car engine, or a machine tool, there may be several stationary regimes, some of which represent in-process states and others that represent out-of-process states. The detection of a change in stationarity can be a possible indication of an out-of-process state, and classification of the new regime can help determine this, and whether additional steps need to be taken. In the area of control, a plant may have several stationary regimes, each of which requires a specially tuned controller (Narendra and Balakrishnan 1997). Detection of a change in the plant and classification of the new regime can be used to switch to the appropriate controller. In time series prediction, detection and classification of a new regime can be used to switch to another more appropriate predictive model.

4.1.1 Contrast with Other Approaches

The mixture model approach has several advantages. By isolating the modeling from the segmentation, it does not require global search techniques for potential change points, and thus it is computationally tractable with nonlinear models. By considering the entire time series simultaneously, the mixture model is capable of recognizing recurring regimes without the expense of modeling the transitions between those regimes.

One of the difficulties in identifying competing approaches is that the literature is spread across many disciplines, such as signal processing, economics, and statistics. We just mention a few of the most prominent here.

Hidden Markov models (HMM) (Baum 1972) are probably the most widely utilized method today for modeling time series whose statistical properties evolve over time, as evidenced by their wide use in speech recognition (see Rabiner and Juang 1993). They postulate a set of hidden states, representing stationary regimes, to each of which is associated a probability density function (pdf) and a matrix of transition probabilities between the states. Generally, HMMs model independent "feature vectors" that describe a locally stationary block of data, and thus the precision with which they can segment is limited by the size of the analysis window. However, in the signal-processing community there are versions, called the hidden Markov filter (Poritz 1982) and mixture autoregressive hidden Markov model (Juang and Rabiner 1985), that model the switching between linear autoregressive systems as a Markov chain and are capable of segmenting at the sample level. There is in fact a very close relationship between these models and the ones discussed in Section 4.9.4, which all fall into the rather broad category of segmental HMMs. The cost of modeling the transitions between regimes is the requirement of large amounts of training data; to segment a single realization of a time series, each regime must be visited multiple times in order to reliably estimate the state transition probabilities or duration distributions.

Switching regression models, popular in the economics community, also propose a hidden state that controls which of several regressors is active. Sometimes the hidden state is viewed as a latent variable that is itself a function of a time series (not necessarily the same one that is being modeled), the simplest example of which is the threshold autoregressive (TAR) model (Tong and Lim 1980). Sometimes the hidden state is regarded as the state sequence of a Markov chain (Goldfeld and Quandt 1973; Hamilton 1989), which again bears a close relationship to some of the models discussed in Section 4.9.4. There are also hybrid versions combining these ideas. Generally, these models are more orientated toward modeling and forecasting rather than segmentation.

Finally, there is the generalized likelihood ratio (GLR) approach, which is a sequential change detection algorithm that does a global search for potential change points in an increasing window of time. Being sequential in nature, the algorithm essentially discards older modeling information and cannot readily recognize recurring regimes. Although the algorithm can theoretically accommodate nonlinear models, it is computationally expensive to do so. We will describe this algorithm in greater detail in Section 4.3 to help us introduce in a graded fashion the two fundamental problems of segmentation and modeling. Moreover, it is a benchmark against which all other segmentation algorithms should be judged.

4.1.2 Contrast with Other Problems

The segmentation problem differs from the so-called blind source separation problem. There, the problem is one of unmixing multiple signals that have been *simultaneously* mixed to produce a new signal. If the component signals are Gaussian random processes, the difference between these two problems is significant. If Gaussian processes are simultaneously added, the resulting process will also be Gaussian, and it is impossible to unmix the signals. Thus, the solution of the blind source separation problem is related to the degree of non-Gaussianity of the signals. As we shall see, this is not the case for the segmentation problem. If several Gaussian processes are switched, the overall process will be multimodal Gaussian and the segmentation problem is still solvable.

The segmentation problem also differs from the one of estimating the parameters of a single system, usually linear, driven by stationary but multimodal noise (Verbout et al. 1998). In this case the presence of multimodality does not imply nonstationarity.

4.1.3   Outline

In the remainder of this section, we define some of the terms and concepts laid out in the goal. In Section 4.2 we review the classical algorithms for supervised segmentation when models are known *a priori*. This assumption solves one of the fundamental issues (the identification) and allows us to treat the case of change detection in a simplified setting. We will later marry these supervised algorithms with unsupervised mixture modeling in Sections 4.4 and 4.5. In Section 4.3 we review unsupervised sequential approaches to segmentation and modeling for the case when little or nothing is known of the data. In Sections 4.4 and 4.5, we show how unsupervised mixture modeling can be combined with the classical supervised segmentation algorithms to produce a new hybrid algorithm.

For the remainder of the chapter we delve into experimental algorithms that are no longer likelihood based. They are derived by analogy with the working principles of the mixture model. They possess both short-term memory sufficient for modeling and long-term memory sufficient for segmentation, and thus attempt to segment and model concurrently during training. These algorithms employ a long-term memory control that is either annealed or adapted during training to effect the granularity of segmentation. Specifically, in Section 4.6, we generalize the discussion to a framework that we call gated competitive experts, under which fall all the new algorithms presented in this chapter, as well as some recent ones in the literature. In Section 4.7 we develop gated competitive principal component analysis. In Section 4.8 we look at various ways of adding memory to gated competitive systems for time-series segmentation without recourse to the classical supervised segmentation algorithms and look at three algorithms. In Section 4.9 we briefly look at some algorithms that attempt to model the transitions between regimes. Finally, in 4.10 we draw conclusions on the present work and suggest possible lines of inquiry for future work. Throughout the chapter, we will present simulations on two benchmark data sets.

At the risk of boring the more advanced reader, we always strive to fully understand simple models first, such as memoryless processes and linear systems. We do so in order to make clear the assumptions required as we progress toward nonlinear systems. We also emphasize simplicity and clarity over rigor and make approximations as early in the treatment as possible. This is a necessity when bringing together so much material in so short a space, but it provides for a unique perspective. Our hope is that this chapter will be accessible to the nonspecialist while perhaps inspiring some new lines of research among specialists.

4.1.4 Stationarity

Khinchin (1932) first defined a stationary random process, and Wold (1938) extended the definition to discrete-time processes. Roughly speaking, a stationary random process is one whose local statistics are invariant with respect to absolute time. In particular, its distributions, and hence its moments, are invariant with respect to the time the random process began, or how long it has been "running," assuming all transients have died out. Such a definition is very broad, so let us break it down into degrees of stationarity. This partitioning is very important, for later on we will find that there is no single test for stationarity but rather multiple tests that correspond to the various degrees of stationary.

Let us first review the concept of a discrete-time random process. Imagine sampling some observable property of a system over a period of time, so that we have a series of measurements $x(n), n = 1, \ldots, N$. Now, imagine that somehow we can reset the system and repeat the measurements. If the results are different, then the process is random (or random with a deterministic component). If we repeat these measurements an infinite number of times, to form an *ensemble*, we can find the first-order probability density function (pdf) of the signal at each sampled time instant: $p(x(n))$. Now, $x(n)$ can be regarded as a random variable. A *first-order stationary process* is one whose first-order pdf is the same for all given sampled times: $p(x(m)) = p(x(n)) \; \forall m, n$. Note that the term *first-order stationarity* refers to the number of random variables only; if the pdf's are the same, all higher order moments must also be the same: $E[x^a(m)] = E[x^a(n)]$ $\forall m, n$.

First-order stationarity does not, however, say anything about relationships *across* time. For that we need the second-order or joint pdf between any two sampled time instants: $p(x(m), x(n))$. A *second-order stationary process* is one where the joint pdf between two sampled time instants depends only on the time difference between them. That is, the joint pdf remains the same for all constant time differences: $p(x(m), x(n)) = p(x(i), x(j))$ when $m - n = i - j$. Second-order stationarity implies that all joint moments are only a function of the time difference: $E[x^a(m)x^b(n)] = f(m - n, a, b)$ $\forall m, n$. The most common joint moment is the autocorrelation function: $r(m, n) = E[x(m)x(n)]$.

Strict stationarity requires that all higher order pdf's be invariant to changes in absolute time. However, estimating multivariate pdf's from data is subject to the so-called dimensionality explosion, requiring an exponentially increasing amount of data with each increase in pdf order, thus rendering impractical higher order stationarity tests. In addition, our interest often lies in the moments themselves, which can be estimated directly from the data, thus bypassing the problem of estimating the pdf's. For these reasons, a lesser requirement known as *wide sense stationarity* is often employed, for which it is only necessary that the first moments of the first- and second-order pdf's be independent of absolute time: $E[x(m)] = E[x(n)]$ and $E[x(m)x(n)] = r(m - n)$ $\forall m, n$. Note that the latter equation implies that the second moment of the first-order pdf is also independent of time: $E[x^2(m)] = E[x^2(n)]$ $\forall m, n$. This has important implications for designing stationarity tests. In particular, tests for a change in autocorrelation or spectrum are also sensitive to changes in power and must be countered by explicit design considerations. Finally, because Gaussian processes are completely characterized by their first and second moments, a wide sense stationary Gaussian process is also strictly stationary.

There is a subset of stationary processes called *cyclo-stationary* for which certain local statistical properties are periodic. Perhaps the best example of such a process is an AR filter excited by a pulse train, which often serves as a production model for voiced speech. Such a signal can be considered strictly stationary by treating the phase of the excitation as random but uniformly distributed within its fundamental period. While the ensemble mean of such a process is constant, the autocorrelation function is periodic.

4.1.5 Ergodicity

There is another, more fundamental problem with these definitions of stationarity. They often cannot be tested in practice because many physical systems are uncontrolled free-running systems that cannot be reset, and thus there exists only one realization of the random process. We then have to hope that the time-average statistics of the single realization are equivalent to the process-average statistics, in which case the process is called *ergodic*. However, estimating the parameters of a cyclo-stationary processes from time averages presents special problems, as valid estimates can only be obtained over integral periods. Note that an ergodic process is stationary but a stationary process is not necessarily ergodic. As for stationarity, there are also different degrees of ergodicity, but we shall not elaborate. We will always assume ergodicity of a random process throughout the following development.

4.1.6 Breakdown of Stationarity

Can all stationary processes be represented by linear systems excited by random noise? The answer is no, as Priestley (1988) has amply demonstrated by counterexample. Some nonlinear systems can produce stationary signals, depending on the type of nonlinearity and the distribution of the excitation. There is, however, a class of signals called *chaotic* for which the very concept of stationarity breaks down. Although there are several properties of chaotic signals, the one of particular importance here is the extreme sensitivity of future values to initial conditions. A nonchaotic system, whether linear or nonlinear, can only be considered stationary after all transients have died out. However, for chaotic systems, the effect of transients never dissipates. For example, consider the joint pdf of a chaotic signal between times m and n, $n > m$, expanded using Bayes' theorem: $p(x(m), x(n)) = p(x(m)) \, p(x(n)|x(m))$. As the time difference becomes large such that $n \gg m$, the conditional pdf $p(x(n)|x(m))$ becomes impossible to determine because a small error in measuring $x(m)$ will result in a vastly different distribution at $x(n)$. Roughly speaking, the approximate time scale over which stationarity breaks down is governed by the Lyapunov exponent, which measures the rate of divergence of neighboring trajectories in state space.

While strict second-order or higher stationarity is ill-defined for chaotic processes, first-order stationarity is in fact defined (see Nicolis, 1995). Certain chaotic processes, called ergodic, evolve toward a stationary first-order distribution, $p(x)$, independent of initial conditions. A general property of ergodic chaotic processes is that they tend not to get "trapped" in restricted regions of phase space for extended periods of time. Time averages thus become equivalent to ensemble averages, hence the similarity with the definition of first-order ergodicity for random processes.

Is it possible to find a new definition of higher order stationarity for chaotic processes that is workable but still maintains similarities with the classical definition? Some have attempted to extrapolate the correspondence between stationarity and constant parameters for linear systems to define stationarity as the output of a dynamical system with constant parameters. However, there are several problems with this definition. First, even for linear systems this correspondence is weak because there may be multiple parameter sets that result in the same multivariate (across time) statistics. For example, a linear system driven by Gaussian noise has a multivariate pdf that is also Gaussian with a covariance matrix whose components are given by the autocorrelation function. Thus, strict stationarity is equivalent to constancy (temporal invariance) of the autocorrelation function or, through the Wiener–Khinchin theorem, the power spectrum. However, there can be many parameter sets that result in the same power spectrum, differing only in phase, and which are thus statistically indistinguishable. It is true that these parameter sets represent distinct discrete points in weight space, and any *incremental* change from one of these points will result in a nonstationary signal. However, it is possible to switch among these different parameter sets in such a way as to maintain a stationary signal. A more serious problem with this definition of stationarity is that the scope of a dynamical system can be expanded to effectively make any process stationary. Both these points will become clearer when we talk about switching models in the context of piecewise stationarity.

How, then, can we define stationarity for chaotic processes? Our approach is to use the classical definition of statistical stationarity as a starting point for developing algorithms for segmentation and identification, all the while seeking opportunities for applying them to nonlinear systems. In the process we will find that many of the algorithms are in fact applicable to chaotic processes because they are based on single-step prediction, for which chaotic processes are indistinguishable from other nonlinear processes. In essence, we arrive at a *working definition* of stationarity that depends on the power of our models to represent parts of the dynamics.

4.1.7 Piecewise Stationarity

Having defined stationarity, we now define piecewise stationarity. Generally, there are two paradigms that can be used to describe piecewise stationary signals. Common to both are a generic state-space representation of a model described by a parameter set θ and an internal state z. There may also be an external excitation source, $u(n)$. A change in either the parameter set or the excitation source may create a new stationary signal. The first paradigm we call parameter *switching* and is diagrammed in Fig. 4.2a. At the transition time, one parameter set or excitation is simply replaced by another while *keeping the state intact*. The second paradigm is called *external switching* and is diagrammed in Fig. 4.2b. At a well-

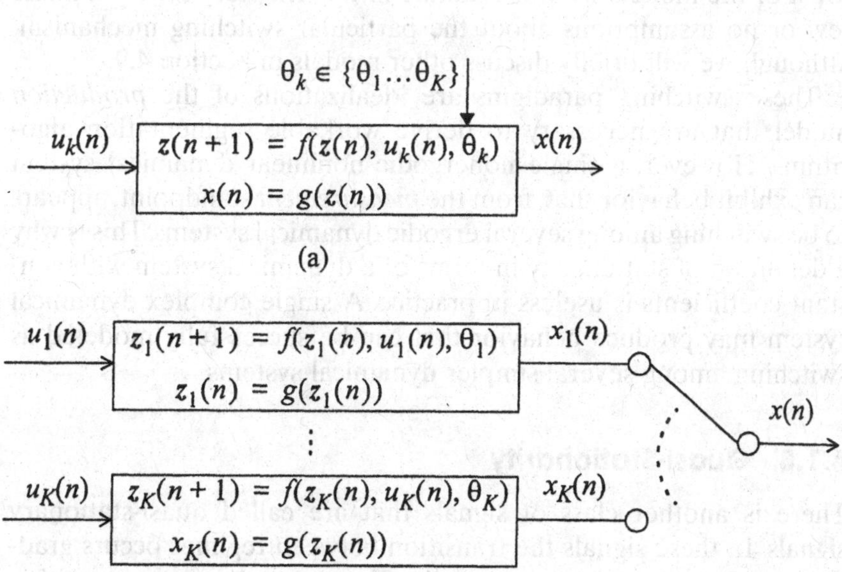

$$\theta_k \in \{\theta_1 \ldots \theta_K\}$$

$$u_k(n) \rightarrow \boxed{\begin{array}{c} z(n+1) = f(z(n), u_k(n), \theta_k) \\ x(n) = g(z(n)) \end{array}} \xrightarrow{x(n)}$$

(a)

$$u_1(n) \rightarrow \boxed{\begin{array}{c} z_1(n+1) = f(z_1(n), u_1(n), \theta_1) \\ z_1(n) = g(z_1(n)) \end{array}} \xrightarrow{x_1(n)}$$

$$\vdots$$

$$x(n)$$

$$u_K(n) \rightarrow \boxed{\begin{array}{c} z_K(n+1) = f(z_K(n), u_K(n), \theta_K) \\ x_K(n) = g(z_K(n)) \end{array}} \xrightarrow{x_K(n)}$$

(b)

Figure 4.2 Piecewise stationary models: (*a*) parameter switching and (*b*) external switching.

defined transition time, the switch connects to a different model, producing a new stationary regime. The primary difference between these two switching paradigms are the transient effects. If the models have memory, then the parameter switching paradigm will exhibit transient effects for the effective memory depth of the system while the external switching paradigm will not exhibit any transient effects. There are differences in the way these two production paradigms are treated statistically, but they are slight as long as the average switching period is greater than the typical memory depth of the systems.

The switching process itself can be modeled in a number of ways, characterized by whether the switch positions and switching times are deterministic or stochastic processes. For example, both the switch position and switching times may be a deterministic function of time. The switch position can also be a deterministic function of the past outputs of the system (the position in state space), in which case the switching times are an implicit function of the state-space trajectory. Alternatively, both the switch positions and switching times can be stochastic, which is usually modeled by assuming the switch position corresponds to the state sequence of a Markov chain. Most of the models we will examine in this chapter, however, make few or no assumptions about the particular switching mechanism, although we will briefly discuss other models in Section 4.9.

These switching paradigms are idealizations of the *production* model that are necessary to derive workable segmentation algorithms. However, a single nonergodic nonlinear dynamical system can exhibit behavior that, from the *measurement* standpoint, appears to be switching among several ergodic dynamical systems. This is why a definition of stationarity in terms of a dynamical system with constant coefficients is useless in practice. A single complex dynamical system may produce behavior that can be successfully modeled as switching among several simpler dynamical systems.

4.1.8 Quasi-Stationarity

There is another class of signals that are called quasi-stationary signals. In these signals the transition between regimes occurs gradually rather than instantaneously. For example, in a transition between two voiced or unvoiced speech segments the excitation remains nearly the same, but the shape of the vocal tract changes, resulting in a continual change in the parameter set of the corresponding dynamic model.

The external switching model can be altered to produce a quasi-stationary signal by replacing the switch with a mixture that is normally dominated by a single system, except during transitions, which are represented by a gradually changing mixture. The parameter switching model can be altered to allow the dynamic model parameters to gradually evolve over time from one set to another. In fact, an adiabatic transition between different parameter sets will generally exhibit less severe transient effects than a quantum switch. The primary focus of this chapter, however, is the study of abrupt changes in dynamical systems.

4.1.9 Approaches to Piecewise Stationary Signals: Tracking Versus Dividing

The question now becomes, given our goal of segmenting and modeling piecewise stationary signals, how can this be accomplished? When confronted with a nonstationary signal, the standard approach is to employ a single adaptive system, such as a Kalman (1960) filter. An adaptive system has continually adjusting parameters that respond to the local (in time) properties of the signal. In order to be local in time, an adaptive system must have a finite effective memory. That is, it must eventually forget what it has learned in the past. The effective memory of an adaptive system may be governed by a learning rate, for example, in the case of the least-mean-square (LMS) algorithm, or a "forgetting factor," in the case of the exponentially weighted recursive least squares (EWRLS) algorithm. However, this finite memory creates a classic trade-off between the speed of convergence and the misadjustment after convergence. Misadjustment describes the variance of the parameters about their nominal solution. In terms of the segmentation and modeling problem, this translates into a trade-off between the accuracy of the modeling and the delay in detecting a new stationary regime. If the effective memory is too short, the parameters exhibit a large variance about the mean, which can even masquerade as a transition. If the effective memory is too long, transition times become uncertain or worse, short stationary regions may be missed entirely.

There is, however, a more fundamental limitation in using a single adaptive system to monitor a nonstationary signal. The normal procedure for detecting a change involves monitoring the innovations of the adaptive system for deviation from whiteness, or a change in variance, the innovations being the part of the signal that cannot be explained by the model. Unfortunately, it can be shown that the sequence of innovations of a single system is an *insufficient statistic* (see Basseville and Nikiforov 1993). That is, there is a many-to-one mapping from different stationary regimes to identical statistics of the innovations. For this reason, two or more adaptive models are required, and this forms the basis of the sequential segmentation algorithms discussed in Section 4.3. These algorithms are called sequential because they only detect a *change* between two stationary but statistically different adjacent temporal *regions*, and information from prior stationary regions is discarded. That is, they are local in both time and model. Under the production models of Fig. 4.2, such methods are appropriate whenever the potential number of dynamical models is very large, so that there is a small probability of returning to the same model. In this case, it is appropriate to view the data as a sequence of stationary *regions*, and the segmentation problem becomes one of detecting the *transition* between two stationary regions.

There is another, fundamentally different, approach to dealing with piecewise stationarity. Instead of viewing a time series as a sequence of unrelated stationary temporal *regions*, it is modeled under the assumption that it has been produced by switching among a *finite set* of stationary *regimes*. For example, in the quasi-stationary assumption for speech, for any given language there are a finite number of statistically different fundamental speech units, sometimes called phonemes, which in this context we would call regimes. This approach is global in time and local in model but is only feasible when the number of potential regimes is reasonably small. Modeling the regime instead of the region has a number of advantages. First and foremost, if several discontiguous regions belong to the same regime, information from all these regions can be used to improve the model estimate. Second, a segmentation system that models the regimes can generalize to new piecewise stationary processes as long as they are generated by the same set of regimes. The development of various algorithms that segment a signal by modeling the regimes forms the fundamental topic of this chapter, beginning in Section 4.4. However, in the next two sections we first present standard supervised and unsupervised sequential methods for change detection and classification.

4.2 SUPERVISED SEQUENTIAL CHANGE DETECTION

Although we are interested in unsupervised segmentation algorithms, it is important to first understand sequential change detection and classification of an *unknown* time series composed of
switching among a finite set of *known stochastic processes*. This task
is sometimes called the *detection/isolation* problem in the statistics
literature and *segmentation/classification* in the neural network
literature. We will use these terms interchangeably throughout
this chapter. Note that this is different from the more commonly
presented problem of detecting known deterministic signals in an
unknown signal (often in the presence of noise), which in its simplest
formulation naturally leads to a bank of matched filters (a correlation receiver).

There are two basic approaches to stochastic time-series classification. In the first, the time series is locally windowed, and a *time-to-
space* mapping transforms the windowed data into a vector of
features, which are then classified as *static* patterns. This process is
shown in Fig. 4.3. If the transformation is chosen properly, distinct
stationary regions of interest will be separable in the feature space.
From our standpoint, however, there are several problems with this
approach. First, temporal resolution is lost through the windowing
process. This may or may not be problematic in itself, depending on
the desired granularity of change detection. However, if a window
overlaps two different stationary regions, there will be degradation

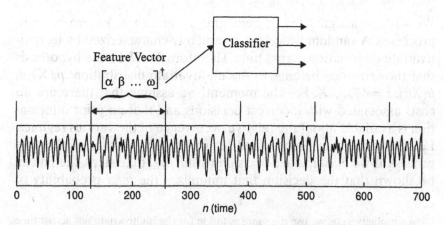

Figure 4.3 Time-series classification utilizing features.

in the feature vector, the severity of which will depend on the transformation. Second, finding a transformation that preserves separability in the feature space is a difficult task and is problem dependent. A common approach is to simply try several standard transformations and use the one that works best.

The approach we present here works directly with the likelihood of the raw time series as measured against probabilistic models for each class of stochastic process. It begins with an initialization step where the active process at the beginning of the time series is identified. This is accomplished by a generalization of the sequential probability ratio test (SPRT) (Wald 1947) to the multiple-hypothesis case (Armitage 1947). The time series is then monitored for a possible change and, if a change is indicated, the new process is identified. This is accomplished by a generalization of the cumulative sum (CUSUM) test (Page 1955) to the multiple-hypothesis case (Zhang 1989). Although the SPRT only serves as a seed for starting the CUSUM test, which is then used to segment the entire time series, we begin with the SPRT because it is easier to understand and many of the principles carry forward to the CUSUM test.

4.2.1 Multiple-Hypothesis SPRT

Let $\mathbf{X}(n)$ represent the history of a single realization (measurement) of a random process over n discrete-time steps

$$\mathbf{X}(n) \equiv [x(n)x(n-1), \ldots, x(1)] \tag{4.1}$$

We wish to assign $\mathbf{X}(n)$ as belonging to one of K known random processes. A random process is completely characterized by its multivariate distribution across time. Therefore, let H_i be the hypothesis that the sequence belongs to the multivariate distribution[1] $p_i(\mathbf{X}) \equiv p(\mathbf{X}|i)$, $i = 1, \ldots, K$. For the moment, we assume that there are no costs associated with incorrect decisions and that no prior information is available about the relative occurrence of the various regimes. Let i^* be the true process and let $\hat{i}^*$ be our estimate of the true process. If we *must* make a decision within n_{max} time steps, then it can be shown that the decision that minimizes the *total* probability of

---

[1] For simplicity sake, we use the same notation for the multivariate pdf across time and the likelihood of the same pdf evaluated at a particular sequence; it should be clear from the context which is intended.

misclassification is to choose the hypothesis $H_{\hat{i}*}$ based on the most likely sequence at time n_{max}

$$\hat{i}* = \operatorname{argmax}_i[p_i(\mathbf{X}(n_{max}))] \tag{4.2}$$

If the goal is fast detection, at the expense of false alarms, then the multiple-hypothesis SPRT is appropriate. The log-likelihood ratio between any two candidate pdf's evaluated at the measured sequence is

$$L_{ij}(n) = \log\left[\frac{p_i(\mathbf{X}(n))}{p_j(\mathbf{X}(n))}\right] \qquad i,j = 1,\dots,K \tag{4.3}$$

where $L_{ii}(n)$ is trivially zero. As we will show later, if $i = i*$ is the true process, from information theory we know that, *on average*, $L_{i*j}(n)$ will be nonnegative and will increase with time for all $j \neq i*$. The multiple-hypothesis SPRT exploits this through a set of parallel Neyman–Pearson tests for deciding among the K hypotheses. That is, a hypothesis $H_{\hat{i}*}$ is chosen if there exists an index $\hat{i}* = i$ for which $L_{ij}(n)$, called the *decision function* in this context, is greater than a set of thresholds

$$L_{ij}(n) \underset{\substack{\text{monitor}}}{\overset{\substack{H_{\hat{i}*=i}\\ \geq}}{}} \beta_{ij} \qquad \forall j \neq i \tag{4.4}$$

The notation in (4.4) is such that if the *stopping rule* passes, testing stops and the upper *decision rule* is applied. Otherwise the lower action is taken, which in this case is continued monitoring. The detection thresholds, β_{ij}, are positive constants.

Although there are $K(K - 1)$ separate likelihood ratios to evaluate, only $K - 1$ of them are actually independent. It follows directly from (4.3) that $L_{ji}(n) = -L_{ij}(n)$. In addition, if L_{ik} is known for all i, the others can be evaluated from

$$L_{ij}(n) = L_{ik}(n) - L_{jk}(n) \tag{4.5}$$

This can be exploited to simplify the test in (4.4) by defining a simplified decision function

$$L_i(n) = \sum_{j=1}^k L_{ij}(n) \tag{4.6}$$

and then choosing the hypothesis $H_{\hat{i}*}$ if *any one* of the L_i exceeds a single predefined threshold, as shown by

$$L_i(n) \underset{\text{monitor}}{\overset{H_{i^*=i}}{\geq}} \beta \quad \exists i \tag{4.7}$$

or, if the test fails, no decision is made and monitoring continues. To show that this test works, first note that from (4.5) and (4.6) we have

$$L_i(n) - L_j(n) = \sum_{k=1}^{K} L_{ik}(n) - \sum_{k=1}^{K} L_{jk}(n) = \sum_{k=1}^{K} L_{ij}(n) = KL_{ij}(n) \tag{4.8}$$

and upon rearranging

$$L_j(n) = L_i(n) - KL_{ij}(n) \tag{4.9}$$

Now, let $i = i^*$ be the true process. Then (4.9) becomes

$$L_j(n) = L_{i^*}(n) - KL_{i^*j}(n) \tag{4.10}$$

Again, information theory tells us that, on average, both $L_{i^*}(n)$ and $L_{i^*j}(n)$ will increase with time so that $L_j(n) < L_{j^*}(n)$ and thus L_{i^*} should be the first to exceed the threshold.

4.2.2 Multiple-Hypothesis CUSUM Test

Having established the active process at the beginning of the time series, the next step is to monitor the time series until *detection* of a statistically significant change and to then *isolate* (classify) the new active process. For the two-hypothesis case, if the data is known to be in one regime and a change is detected, there is only one possibile choice for the new regime. In this case, detection automatically accomplishes classification. However, for the multiple-hypothesis case the situation is much more complex because both a change detector and classifier are required, and they may have competing requirements. For example, while minimizing the delay of the detector is generally desirable, the classifier may be more accurate by using the additional data associated with a longer detection delay. Nikiforov (1995) was the first to consider the *joint* optimal design of the detection and isolation stages. Unfortunately, Nikiforov's algorithm is not recursive and therefore we consider an alternative formulation due to Zhang (1989).

Assume that the current active process has been correctly identified as belonging to $p_{i^*}(X)$, the data arrive one sample at a time, and we wish to determine if and when a change occurs to $p_{j*}(X)$. We know that under $i*$, the process likelihood $L_{i*j}(n)$ will increase with time. However, for historical reasons, the CUSUM test tracks $L_{ji*}(n) = -L_{i*j}(n)$, which decreases with time. The CUSUM test is essentially an SPRT that continually resets the log-likelihood for the active process relative to its minimum value

$$\Delta L_{ji*}(n) = L_{ji*}(n) - \min[L_{ji*}(n-1), L_{ji*}(n-2), \dots L_{ji*}(1)] \qquad \forall j \neq i* \tag{4.11}$$

in order to maintain maximum sensitivity to any potential change. This can also be written recursively as

$$\Delta L_{ji*}(n) = \max[0, \Delta L_{ji*}(n-1) + l_{ji*}(n)] \qquad \forall j \neq i* \tag{4.12}$$

where l_{ji*} is the *log-likelihood increment*, which we shall define shortly. The other values are determined relative to the active process, similar to (4.5):

$$\Delta L_{jk}(n) = \Delta L_{ji*}(n) - \Delta L_{ki*}(n) \tag{4.13}$$

Under $i*$, both $\Delta L_{ji*}(n)$ and $\Delta L_{ki*}(n)$ will be repeatedly reset to zero, causing $\Delta L_{jk}(n)$ to fluctuate around zero. If a change occurs to $j*$, then $\Delta L_{j*k}(n)$ will tend to increase for $\forall k \neq j$. This forms the basis for a test that can be used to decide if a change has occurred, and if so, identify the new active process. A hypothesis H_{j*} is chosen if there exists an index $\hat{j}* = j$ for which the decision function $\Delta L_{jk}(n)$ is greater than a set of thresholds

$$\Delta L_{jk}(n) \underset{\text{monitor}}{\overset{H_{j*} = j}{\gtrless}} \beta_{ik} \qquad \forall k \neq j \tag{4.14}$$

If the test fails, monitoring continues. Once a detection occurs, the change point can be estimated as the time at which L_{i*j*} was maximum:

$$\hat{T} = \arg\max_n [L_{i*j*}(n)] \tag{4.15}$$

The CUSUM test is then reset and started anew from the change point.

Just as for the SPRT, we prefer a simplified decision function

$$\Delta L_j(n) = \sum_{k=1}^{K} \Delta L_{jk}(n) \tag{4.16}$$

that has only a single threshold, such that if any one of the ΔL_j exceeds the threshold,

$$\Delta L_j(n) \underset{\text{monitor}}{\overset{H_{j*=j}}{\geq}} \beta \quad \exists j \tag{4.17}$$

the test stops and a decision is made in favor of a change. Kerestecioglu (1993) has also proposed a modified multiple-hypothesis CUSUM test but where the process log-likelihood ratio differences in (4.12) are reset to zero all together or not at all.

If there are prior probabilities associated with each regime, then, in place of the raw likelihoods in the SPRT and CUSUM, we use the posterior probabilities, $P_i p_i(\mathbf{X}(n))$, where P_i is the prior probability of the ith regime.

4.2.3 Recursive Calculation of the Likelihood Ratio

It would clearly be advantageous to calculate the log-likelihood ratio recursively instead of having to know multivariate distributions that increase in dimensionality at every time step. This is easily accomplished using Bayes' theorem to expand the sequence likelihood

$$p(\mathbf{X}(n)) = p(x(n), \mathbf{X}(n-1)) = p(x(n)|\mathbf{X}(n-1))p(\mathbf{X}(n-1)) \tag{4.18}$$

and then through recursion

$$p(\mathbf{X}(n)) = \prod_{m=1}^{n} p(x(m)|\mathbf{X}(m-1)) \tag{4.19}$$

where it is understood that $p(x(1)|\mathbf{X}(0)) \equiv p(x(1))$. Applying this to the multiple-hypothesis case, the process log-likelihood ratio in (4.3) can now be written

$$L_{ij}(n) = \sum_{m=1}^{n} l_{ij}(m) \tag{4.20}$$

where $l_{ij}(m)$ is the *log-likelihood increment*

$$l_{ij}(n) = \log\left[\frac{p_i(x(n)|\mathbf{X}(n-1))}{p_j(x(n)|\mathbf{X}(n-1))}\right] \tag{4.21}$$

where we have defined $p_i(x(n)|\mathbf{X}(n-1)) \equiv p(x(n)|\mathbf{X}(n-1), i)$. Now the summation in (4.20) admits a recursive calculation:

$$L_{ij}(n) = L_{ij}(n-1) + l_{ij}(n) \tag{4.22}$$

For memoryless random processes, more commonly known as independent identically distributed (i.i.d.) sequences, the current value of the time series is independent of its history, and the log-likelihood increment in (4.21) has the particularly simple form

$$l_{ij}(n) = \log\left[\frac{p_i(x(n))}{p_j(x(n))}\right] \tag{4.23}$$

where $p_i(x)$ is the distribution of the ith i.i.d process.

For nonlinear processes with memory, finding the theoretical multivariate pdf's may be difficult. Or, even if we know the parametric form of the theoretical pdf's, we may only have a short experimentally measured time series for each candidate process from which we cannot reliably determine the parameters of the pdf's. An example of such a practical time-series segmentation/classification problem is shown in Fig. 4.4, which shows three short exemplars representing each of three possible regimes (classes) and an unknown time series composed of occasional switching among the three regimes. Again, we want to emphasize that the problem here is not one of detecting when the exemplars occur deterministically within the unknown time series but rather of determining when statistically similar stochastic processes occur.

Given these practical constraints, we can use prediction as a tool to evaluate the likelihoods in (4.21) directly. If we can find models under each hypothesis that can predict $x(n)$ from prior data, up to a random term

$$x(n) = f_i(\mathbf{X}(n-1)) + e_i(n) \qquad i = 1,\dots,K \tag{4.24}$$

then we can evaluate the conditional probability in terms of the pdf of the *innovations*, $e_i(n)$:

$$p_i(x(n)|\mathbf{X}(n-1)) = p_{e_i}(e_i(n)) \qquad i = 1,\dots,K \tag{4.25}$$

Up to now we have been loosely using the term *nonlinear signal*, but now we are in a position to formally define it: A nonlinear signal is a signal that cannot be *whitened* through the prediction process using a time-invariant linear transformation. By whitened, we mean that the innovations must be truly independent.

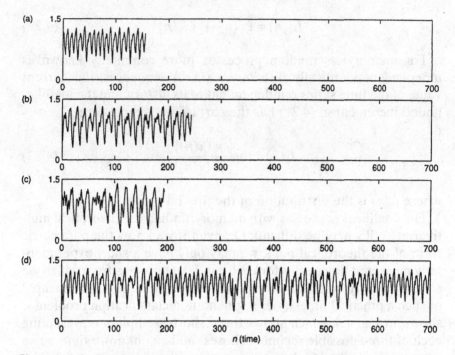

Figure 4.4 Practical time-series segmentation/classification problem: (a–c) time-series exemplars representing each of three different classes and (d) unknown time series consisting of occasional switching among the three classes.

Using (4.25) the process log-likelihood ratios are now given by

$$L_{ij}(n) \approx \sum_{m=1}^{n} l_{ij}(m) \tag{4.26}$$

where the log-likelihood increments are now given in terms of the prediction errors:

$$l_{ij}(n) = \log\left[\frac{p_{e_i}(x(n) - f_i(\mathbf{X}(n-1)))}{p_{e_j}(x(n) - f_j(\mathbf{X}(n-1)))}\right] \tag{4.27}$$

Equation (4.26) is written as an approximation because we have ignored initial conditions and for additional reasons that are discussed below. Once again, this achieves the goal of recursive computation of the process log-likelihood ratios as in (4.22).

This expression of the likelihood of a sequence in terms of prediction error is known as the *prediction error decomposition*. The advantage of the prediction error decomposition is that it allows us to evaluate the likelihood of a sequence in terms of the likelihood of a sequence of prediction errors, without ever having to know the multivariate (across time) pdf of the process. However, prior to evaluating the likelihood of a sequence, we need to have developed a predictive model for the data and a probabilistic model for the innovations. We now discuss some of the issues involved in each.

In practice, the entire past sequence is not needed to predict the next sample. For example, for an AR(p) process, only p past samples are required to predict the next sample and any additional history does not improve the prediction. In terms of the statistics of the random process, the conditional distribution of the random process given its past becomes independent of its distant past. If the *effective memory depth* is M for a given process, then (4.24) and (4.25) become, respectively,

$$p(x(n)|\mathbf{X}(n-1)) \approx p(x(n)|\chi_M(n-1)) \tag{4.28}$$

$$x(n) \approx f(\chi_M(n-1)) + e(n) \tag{4.29}$$

where $\chi_M(n)$ is a vector of the M most recent samples of the process

$$\chi_M(n) \equiv [x(n), x(n-1), \ldots, x(n-M+1)]^\dagger \tag{4.30}$$

which is, in effect, the output of a tapped-delay line with M taps. Such a process is formally called an Mth-order *Markov process*. For ease of notation, we shall just use $\chi(n)$ from this point forward, with the understanding that the number of taps, M, is an implicit design parameter.[2]

In this chapter, when confronted with nonlinear processes with memory, we will use neural network predictors to implement the functional mapping in (4.29). When a neural network is fed by a tapped-delay line it falls into the category of a *time-delay neural network* (TDNN) (Waibel et al. 1989). In keeping with the spirit of

---

[2] To summarize: $x(n)$ is the time series itself, $\chi(n)$ its recent past, and $\mathbf{X}(n)$ its entire history. The advantage of this notation is that it can equally represent scalar or vector time series without modification.

this book, this architecture is entirely feedforward, possessing no recurrent connections, and can be trained using gradient descent techniques.

There is a fundamental trade-off in using prediction to evaluate the process log-likelihood ratio. Since only one predictor at a time can be modeling the true process, the errors of the other predictors will not be the innovations but rather the residuals (the only exception to this is the somewhat trivial case where two linear predictors model the same process at different power levels). This means that the log-likelihood ratio increments will not be independent, which complicates theoretical analysis. Furthermore, in practice we will implement the predictors using neural networks trained on finite time-series exemplars. They may then have to generalize their predictions to regions of signal space over which they were not trained, which is of particular concern for the predictors that do not represent the current true process. We shall have more to say about this at the end of this section. Even for the predictor that currently represents the true process, one must be careful in assuming independence of the innovations. For example, the optimal minimum mean-square-error (MMSE) predictor can be shown to be the conditional expectation

$$f(\mathbf{X}(n-1)) = E[x(n)|\mathbf{X}(n-1)] \qquad (4.31)$$

In this case, the most that can be said about the innovations is that they are uncorrelated, $E[e(m)e(n)] \propto \delta_{m,n}$. If they were not, then one could find a better model that would exploit the correlations and improve the prediction. However, this does not tell us whether the innovations are independent. The exception is if they are Gaussian distributed, in which case uncorrelation implies independence. In spite of all these caveats, the importance of the prediction error decomposition is that it allows us to move from probabilistic to model-based algorithms and opens the door to the possibility of segmenting complex nonlinear signals and even chaotic signals, as we will see shortly. Its ultimate justification is that the resulting algorithms work well in practice.

4.2.4 Examples

For the problem of detecting a change in a memoryless i.i.d. vector Gaussian process, let H_i be the hypothesis that $\mathbf{x}(n)$ is generated according to $N(\boldsymbol{\mu}_i, \Sigma_i)$. The log-likelihood increment is then

$$2l_{ij}(n) = [\mathbf{x}(n) - \boldsymbol{\mu}_j]^\dagger \Sigma_j^{-1}[\mathbf{x}(n) - \boldsymbol{\mu}_j] - [\mathbf{x}(n) - \boldsymbol{\mu}_i]^\dagger$$
$$\Sigma_i^{-1}[\mathbf{x}(n) - \boldsymbol{\mu}_i] + \log\left[\frac{|\Sigma_j|}{|\Sigma_i|}\right] \tag{4.32}$$

The first two terms represent the difference between the Mahalonobis distances of the current sample to the two models.

For the problem of detecting a change in a scalar random process with memory using predictive models under the assumption of Gaussian innovations, let H_i be the hypothesis that the random process can be modeled as $x(n) = f_i(\chi(n-1)) + e_i(n)$, where the residuals are i.i.d. Gaussian $e_i \sim N(0, \sigma_i^2)$. The log-likelihood increment is then

$$2l_{ij}(n) = \frac{e_j^2(n)}{\sigma_j^2} - \frac{e_i^2(n)}{\sigma_i^2} + \log\left[\frac{\sigma_j^2}{\sigma_i^2}\right] \tag{4.33}$$

The first two terms represent the weighted difference between the squared residuals of the two predictive models.

4.2.5 Summary of Practical Supervised Time-Series Segmentation/Classification

Because of the large amount of background material presented, a summary of supervised time-series segmentation and classification is in order. We make the realistic assumption that instead of theoretical pdf's, we are in possession of a set of experimentally measured time series, each of which represents a statistically distinct random process. We first develop models for each of the candidate processes. For memoryless processes, these models are static probability distributions. For processes with memory, we develop a predictive model for each candidate process such that the innovations are independent and a probabilistic model for the innovations.

To detect and classify these candidate processes within an unknown time series, we use the log-likelihood ratio increments between the trained models as measured on the unknown time series. First, the initial active process at the beginning of the time series is classified using the multiple-hypothesis SPRT. Then, the time series is monitored for change to a different process using the multiple-hypothesis CUSUM test. Once a change is detected, the CUSUM test is reset and monitoring continues anew. The whole process of model building and detection/isolation is summarized in Fig. 4.5 for nonmemoryless processes.

4.2.6 Information-Theoretic Interpretation

As eluded to earlier, there is an information-theoretic interpretation to the change detection algorithms. If $i*$ is the true active process, the expected value of the process log-likelihood ratio between two *stationary* processes is

$$E_{i*}[L_{ij}(n)] = \sum_{m=1}^{n} E_{i*}[l_{ij}(m)] = n E_{i*}[l_{ij}] \qquad (4.34)$$

Thus, the expected value of the log-likelihood increment, $E_{i*}[l_{ij}]$, represents the average slope of the process log-likelihood ratio with time. In general, this slope can be either positive or negative. For the particular case when $i = i*$, however, the average slope of the log-likelihood increment

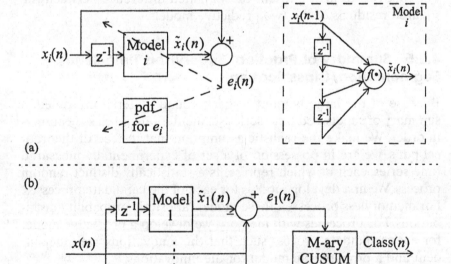

(a)

(b)

Figure 4.5 Segmentation/classification of nonmemoryless time series: (*a*) predictive model building and (*b*) application of the multiple-hypothesis CUSUM algorithm for segmentation/classification of an unknown time series.

$$E_{i*}[L_{i*j}(n)] = nE_{i*}[l_{i*j}] = nK_{i*j} \qquad (4.35)$$

is the asymmetric Kullback–Leibler (KL) divergence, which for memoryless processes (4.23) is

$$K_{i*j} = \int p_{i*}(x) \log\left[\frac{p_{i*}(x)}{p_j(x)}\right] dx \geq 0 \qquad (4.36)$$

For example, for an i.i.d vector Gaussian process with distribution $N(\mu_i, \Sigma_i)$, the expected value of the increment (4.32) results in

$$2K_{i*j} = [\mu_{i*} - \mu_j]^\dagger \Sigma_j^{-1} [\mu_{i*} - \mu_j] + \text{Tr}[\Sigma_{i*}\Sigma_j^{-1}] + \log\left[\frac{|\Sigma_j|}{|\Sigma_{i*}|}\right] - D \qquad (4.37)$$

where D is the dimension of the vector.

Since the KL divergence is always nonnegative for any two different pdf's, the expected value of the process log-likelihood $L_{i*j}(n)$ *prior* to change increases with time at a rate determined by the KL divergence K_{i*j}. Likewise, if $j*$ is the true active process *after* the change point, the expected value of the process log-likelihood $L_{ij*}(n)$ decreases with time at a rate determined by K_{ij*}, which is a negative quantity for all i.

For stationary random processes with finite memory, we can use the prediction error decomposition to evaluate the expected value of the log-likelihood increment in terms of the residuals

$$F_{i*}[l_{i*j}] = F_{i*}\left[\log\left[\frac{p_{e_{i*}}(e_{i*})}{p_{e_j}(e_j)}\right]\right] = K_{i*j} \qquad (4.38)$$

where it can be shown that K_{i*j} is equal to the Kullback–Leibler divergence between two random processes, which is formally defined in terms of the multivariate pdf's across time:

$$K_{i*j} = \lim_{n \to \infty} \frac{1}{n} \int p_{i*}(\mathbf{X}(n)) \log\left[\frac{p_{i*}(\mathbf{X}(n))}{p_i(\mathbf{X}(n))}\right] d\mathbf{X}(n) \geq 0 \qquad (4.39)$$

Therefore, everything said about memoryless random processes also applies here, except now the expected slope of $L_{i*j}(n)$ in (4.34) is governed by the multivariate (across time) KL divergence.

If predictive models are known, it is easier to evaluate the KL divergence directly in terms of the pdf's of the innovations (4.38). For example, for the special case of Gaussian innovations, the expected value of (4.33) becomes, by inspection,

$$2K_{i*j} = \frac{1}{\sigma_j^2} E_{i*}[e_j^2(n)] - \log\left[\frac{\sigma_{i*}^2}{\sigma_j^2}\right] - 1 \qquad (4.40)$$

For Gaussian autoregressive moving average (ARMA) processes, the KL divergence is conveniently expressed in terms of the power spectra of the processes

$$2K_{i*j} = \frac{1}{2\pi} \int_{-\pi}^{\pi} \left[\frac{S_{i*}(f)}{S_j(f)} - \log\left[\frac{S_{i*}(f)}{S_j(f)}\right] - 1\right] df \qquad (4.41)$$

where $S(f)$ is the power spectra evaluated at the digital frequency f. The correspondence between the first terms of (4.40) and (4.41) follows directly from Parseval's relation, while the second terms are related through the Kolmogoroff–Szegö formula (see Sugimoto and Wada 1988). It is worthwhile noting that the Itakura–Saito spectral distance is 4π times the KL divergence (Pinsker 1964). For Gaussian AR processes, the KL divergence can be expressed in closed form:

$$2K_{i*j} = \frac{1}{\sigma_j^2}(\mathbf{w}_{i*} - \mathbf{w}_j)^\dagger \mathbf{R}_{i*}(\mathbf{w}_{i*} - \mathbf{w}_j) + \frac{\sigma_{i*}^2}{\sigma_j^2} - \log\left[\frac{\sigma_{i*}^2}{\sigma_j^2}\right] - 1 \quad (4.42)$$

where $\mathbf{w}$ is a vector of linear prediction coefficients, $\mathbf{R}$ is the process Toeplitz autocorrelation matrix, and σ^2 is the power of the noise driving the AR processes.

The first term in (4.40) can be interpreted as the normalized mean-square *cross-prediction* error of the jth predictive model when the data is generated according to $i*$. Using the relative frequency interpretation of expectation, this interpretation allows the KL divergence to be measured experimentally by feeding the data of the true process $i*$ into the predictive model for process j. Assuming ergodicity, the expected value of the squared prediction error can then be estimated by a time average. Figure 4.6a shows the model-building stage of this two-step process, while Fig. 4.6b shows the measurement of the cross-prediction error variance. In this way, the KL divergence can be estimated between two nonlinear processes with Gaussian

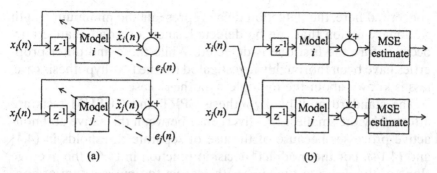

Figure 4.6 Experimental KL divergence measurement for processes with memory under a Gaussian assumption on the residuals: (*a*) predictive model building and (*b*) measurement of cross-prediction error variance.

innovations for which there is no known analytical expression for the multivariate (across time) pdf's. If the innovations are non-Gaussian, $E_{i*}[l_{i*j}]$ can still be evaluated in terms of the innovations of the predictive models, but the measurement setup in Fig. 4.6*b* will change accordingly. Experimental cross-prediction error variance has previously been proposed as a general distance measure (Schreiber 1997) but without the scaling factors in (4.40) and without recognition of the close relationship to the KL divergence.

4.2.7 Detectability of Change

We now turn to the question of the *detectability* of change for the CUSUM test. By detectability, we mean both the false alarm rate before the change point and the detection delay after the change point. In the literature, the mean false alarm rate is called the operating characteristic (OC) while the mean detection delay is called the average sample number (ASN). For the two-hypothesis case (Lorden 1971), it has been shown that in the limit of a large threshold, the CUSUM minimizes the mean detection delay for a fixed false alarm rate. Because the decision function is continually reset to zero before the change, the mean false alarm rate is independent of K_{i*j} but inversely proportional to the threshold. After the change, the mean detection delay is proportional to the log of the threshold and inversely proportional to K_{ij*}, as would be expected geometrically given the interpretation of K_{ij*} as the slope of $L_{ij*}(n)$ after the change. For the off-line segmentation problem, of which we are primarily concerned here, the detection delay represents the minimum length stationary region that can be detected, since a second change may occur before the first one is detected. While these and other properties have been thoroughly investigated for the two-hypothesis case, less is known about the multiple-hypothesis case.

The standard multiple-hypothesis SPRT and CUSUM tests are affected by the smallest KL divergence between the active and nonactive processes because of the use of separate thresholds in (4.4) and (4.14). For the modified decision function in (4.6), the average slope of the decision function with respect to time is a sum of nonnegative KL divergence measures

$$E_{i*}[L_{i*}(n)] = n \sum_{j=1}^{K} K_{i*j} \qquad (4.43)$$

which is dominated by the *largest* KL divergence. The single threshold in (4.6) and (4.16) is an important simplification because the false alarm rate is often a fundamental design constraint.

Because the KL divergence is not necessarily symmetric, detecting a change from process A to process B may be more or less difficult than detecting a change from B to A. For example, in the case of detecting a change in the variance of an i.i.d. process, the nonsymmetry of the KL divergence results in it being easier to detect an increase in variance compared to a decrease in variance. For this reason, various modifications to the log-likelihood increment have been proposed (see Basseville and Nikiforov 1993). In the divergence algorithm (Basseville and Benveniste 1983), a term is added to the log-likelihood increment such that its expected value leads to the *symmetric* Kullback–Leibler divergence, where the symmetry is with respect to an interchange of the true processes, $i*$ and $j*$, before and after change. In fact, the log-likelihood increment is just one of a number of possible increments that can be used, each of which will lead to a different distance metric under the expectation. Kazakos and Papantoni-Kazakos (1990) provide a thorough summary of various probabilistic distance metrics and their properties.

4.2.8 Other Approaches

Rather than working with likelihood ratios, it may be more intuitive to work directly with probabilities. Using Bayes's theorem, the probability of the ith model given the data up to the current time n can be written as

$$P(i|\mathbf{X}(n)) = \frac{p(\mathbf{X}(n)|i)}{\sum_{k=1}^{K} p(\mathbf{X}(n)|k)} \equiv \frac{p_i(\mathbf{X}(n))}{\sum_{k=1}^{K} p_k(\mathbf{X}(n))} \qquad i = 1, \ldots, K \quad (4.44)$$

or

$$P(i|\mathbf{X}(n)) = \frac{\prod_{m=1}^{n} p_i(x(m)|\mathbf{X}(m-1))}{\sum_{k=1}^{K} \prod_{m=1}^{n} p_k(x(m)|\mathbf{X}(m-1))} \qquad (4.45)$$

This also admits a recursive computation:

$$P(i|\mathbf{X}(n)) = \frac{P(i|\mathbf{X}(n-1)) p_i(x(n)|\mathbf{X}(n-1))}{\sum_{k=1}^{K} P(k|\mathbf{X}(n-1)) p_k(x(n)|\mathbf{X}(n-1))} \qquad (4.46)$$

which has the nice interpretation that the posterior model probabilities at the previous time step become the prior probabilities for the current time. For processes with memory, if we make the same assumptions as in Section 4.23, namely that the signals have finite memory and that the class conditional pdf's can be evaluated using predictive models, then (4.46) can be written

$$P(i|\mathbf{X}(n)) = \frac{P(i|\mathbf{X}(n-1)) p_{e_i}(x(n) - f_i(\boldsymbol{\chi}(n-1)))}{\sum_{k=1}^{K} P(k|\mathbf{X}(n-1)) p_{e_k}(x(n) - f_k(\boldsymbol{\chi}(n-1)))} \qquad (4.47)$$

This formulation appears in the control literature as the multiple model algorithm for model selection (Bar-Shalom and Fortmann 1988) and in the neural network literature for time-series classification (Petridis and Kehagias 1996; Plataniotis et al. 1996). However, note that the log-likelihood ratio of the probability of modes i and j in (4.46) is

$$\log\left[\frac{P(i|\mathbf{X}(n))}{P(j|\mathbf{X}(n))}\right] = \log\left[\frac{P(i|\mathbf{X}(n-1))}{P(j|\mathbf{X}(n-1))}\right] + \log\left[\frac{p_i(x(n)|\mathbf{X}(n-1))}{p_j(x(n)|\mathbf{X}(n-1))}\right] \qquad (4.48)$$

which is exactly the recursive calculation of the cumulative log-likelihood ratio in (4.22) and (4.21), as used by the multiple-hypothesis SPRT and CUSUM, demonstrating that the two approaches are essentially using the same information. The problem with using (4.46) alone is that it represents the probability of the models given all prior data, rather than the data since the last change point. By not specifying stopping and decision rules, the normalized cumulative likelihood in (4.46) rapidly decreases to zero for out-of-regime predictors ($i \neq i^*$), resulting in a slow response to a change in regime and numerical precision problems.

For the special case of a memoryless process, Baum and Veeravalli (1994) have independently specified appropriate stopping and decision rules for a multiple-hypothesis SPRT using the normalized cumulative likelihood as a decision function, which for the memoryless case from (4.46) is

$$P(i|\mathbf{X}(n)) = \frac{P(i|\mathbf{X}(n-1))p_i(x(n))}{\sum\limits_{k=1}^{K} P(k|\mathbf{X}(n-1))p_k(x(n))} \tag{4.49}$$

A decision is made in favor of the hypothesis H_{i^*} if *any one* of the probabilities exceeds a predefined threshold:

$$P(i|\mathbf{X}(n)) \underset{\text{monitor}}{\overset{H_{i^*=i}}{\geq}} \beta_i \qquad \exists i \tag{4.50}$$

The thresholds, β_i, have the nice property that they are themselves probabilities and hence can be viewed as confidence levels for detection. Although theoretical analysis is less straightforward [the expected value of the decision function (4.49) does not directly yield the KL divergences] the thresholds have been asymptotically related to the OC and ASN. Extension of their results to the memory case using (4.46) as the decision function is straightforward, but we did not employ this SPRT because there is not yet, to the best of our knowledge, a CUSUM-like test for detecting a change to a new regime given knowledge of the current regime.

However, based on analogy with the relationship between the two-hypothesis SPRT and CUSUM, we can postulate a reasonable multiple-hypothesis CUSUM test based on probabilities. As before, assume we have correctly identified the current process as i^* and we wish to determine if and when a change occurs to j^*. The basic idea is to calculate the posterior model probabilities recursively as before. However, before using them as the prior probabilities at the next time step, they are first checked to see if the current process is the most probable. If it is, then they are all reset to equal probability in order to minimize any future detection delay. If the current process is not the most probable, then the posterior probabilities become the prior probabilities without modification in order to "wait and see" whether any new processes become predominate. If we use separate symbols, P and Q, for the posterior and prior model probabilities, respectively, then the procedure can be codified into the following two steps:

$$P_i(n) = \frac{Q_i(n-1)p_i(x(n)|\mathbf{X}(n-1))}{\sum\limits_{k=1}^{K} Q_k(n-1)p_k(x(n)|\mathbf{X}(n-1))} \tag{4.51}$$

$$Q_i(n) = \begin{cases} \dfrac{1}{K} & P_{i*}(n) > P_i(n) \quad \forall i \neq i^* \\ P_i(n) & \text{otherwise} \end{cases} \tag{4.52}$$

The prior probabilities also become the decision function. A change is deemed to occcur if any one of the prior probabilities exceeds a predefined threshold:

$$Q_j(n) \underset{\text{monitor}}{\overset{H_{j*=j}}{\geq}} \beta_j \quad \exists j \tag{4.53}$$

Another way of overcoming the difficulties of using (4.46) directly, without testing for a change point, is to calculate the probability of the various models using only the data from the recent past. For this approach to be valid, the temporal scope of the window of past data used should be less than the average regime switching period, in order to guarantee that the probabilities reflect a single stationary regime. This idea is central to the algorithms presented in Section 4.8.

Kehagias and Petridis (1997a) and Plataniotis et al. (1996) also describe other non-Bayesian variations of (4.46) that we will not discuss here.

These alternative approaches, as well as the CUSUM test, are *nondiscriminative* in the sense that the models are trained *separately* on *a priori* labeled data. In *discriminative* classification, the training and performance of one model can affect the training of another model. Alternatively, in place of multiple models, a single classifier may be employed, an example of which was shown in Fig. 4.3 for feature vector classification. Discriminative training can lead to a more compact representation of the classification process, since the classifier need only be concerned with the *boundaries* between classes. There are discriminative approaches to direct time-series classification (without intermediate feature vectors), but the training task invariably leads to the so-called temporal credit assignment problem, which in this context describes the problem of designing a desired output for a classifier at intermediate stages of a time series. In theory, a finite-memory version of (4.46) could be used to help design a desired output for a direct classifier, but to the best of our knowledge this has not been explored.

4.2.9 Data Sets

We now introduce the data sets that will be used to test the various algorithms as they appear throughout this chapter. Data set A, which also appears as a benchmark in Pawelzik et al. (1996) and Kehagias and Petridis (1997b), consists of switching between four ergodic chaotic time series generated by the logistic map (regime 1)

$$x(n+1) = f(x(n)) = 4x(n)(1 - x(n)) \qquad (4.54)$$

the tent map (regime 2)

$$x(n+1) = g(x(n)) = 1 - 2\,\text{abs}[x(n) - 0.5] \qquad (4.55)$$

the double logistic map (regime 3)

$$x(n+1) = f(f(x(n))) \qquad (4.56)$$

and the double tent map (regime 4)

$$x(n+1) = g(g(x(n))) \qquad (4.57)$$

Since these maps generate ergodic time series, their first-order distributions are well defined. However, due to the application of the double maps, regimes 1 and 3 as well as regimes 2 and 4 possess identical first-order pdf's and thus cannot be distinguished solely on the basis of first-order statistics.

The regimes were switched to cover every possible transition, 1–2–1–3–1–4–2–3–2–4–3–4–1, and each segment had a length that was uniformly random between 50 and 150 samples. Both training and testing time series were created and then truncated to 1300 samples for convenience. Zero-mean Gaussian noise with a standard deviation of 0.02 was added to the final time series to simulate measurement noise (the noise was not part of the dynamics). They are shown in Fig. 4.7, along with their regime truth.

In order to measure the distance between the regimes using the experimental cross-prediction approach of Fig. 4.6, we first trained four neural network predictors separately on the corresponding *labeled* training data. Knowing *a priori* that a single past sample is needed to predict the next sample, we used an embedding dimension of $M = 1$. Each neural network was a multilayer perceptron (MLP) with five hyperbolic tangent activation functions on the single hidden layer, and a single linear output for a total architecture of 1–5–1. This was the minimum size network that was capable of *separately* learning each of the true return maps. This is also the architecture used by all other algorithms for data set A. The return maps

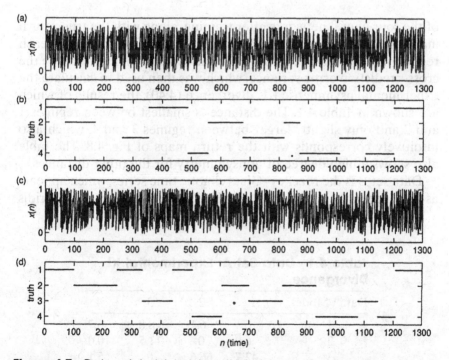

Figure 4.7 Data set A: (*a*) training, (*b*) training truth, (*c*) testing, and (*d*) testing truth.

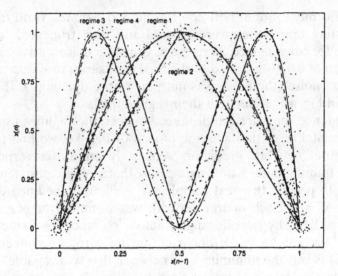

Figure 4.8 Data set A: true return map (solid line), training data (dot), and return map of MLP predictors (dotted line).

of the four neural networks after training, along with the true return maps and training data are shown in Fig. 4.8. The data from each regime was then passed through each model in order to measure the cross-prediction error variance, which was then used to calculate the experimental asymmetric KL divergence (4.40), the results of which are shown in Table 4.1. The distance is smallest between regimes 1 and 2 and only slightly larger between regimes 3 and 4, which also intuitively corresponds with the return maps of Fig. 4.8. The table also shows a high degree of nonsymmetry for the other distances.

Data set B is the Mackey–Glass chaotic time series, which appears as a benchmark in Pawelzik et al. (1996), Kehagias and Petridis

Table 4.1 Data Set A: Experimental KL Divergence

Data \ Model	1	2	3	4
1	0	6.1	8.8	6.7
2	6.0	0	11.5	10.0
3	42.7	66.6	0	5.1
4	28.2	40.1	2.8	0

(1997b), and Ramamurti and Ghosh (1999). It is generated by the first-order differential equation

$$\dot{x}(t) = -0.1x(t) + \frac{0.2x(t-t_d)}{1+x^{10}(t-t_d)} \tag{4.58}$$

where t_d is a time-delay parameter that switches between the values 17 (regime 1), 23 (regime 2), and 30 (regime 3). We generated the discrete time series through fourth-order Runge–Kutta integration, followed by 6:1 downsampling. The parameters were switched during integration, as in Fig. 4.2a, to create transient effects. Separate histograms of the three time series demonstrates that the first-order distributions are very similar but not identical. In particular, a larger delay parameter creates a broader distribution.

The parameters were switched to cover every possible transition, 1–2–1–3–2–3–1, and each segment had a length that was uniformly random between 50 and 150 samples. Both training and testing time series were created and truncated to 700 samples for convenience. Zero-mean Gaussian noise with a standard deviation of 0.02 was added to the final time series to simulate measurement noise (the noise was not part of the dynamics). They are shown in Fig. 4.9 along with their regime truth.

We trained three neural network predictors separately on the corresponding *labeled* training data. Based both on the literature and experiment, we chose an embedding dimension of $M = 6$. The neural networks were multilayer perceptrons with six hyperbolic tangent activations functions on the single hidden layer, and a single linear output for a total architecture of 6–6–1. This was the minimum size network that provided comparable results on all three regimes, as determined by a scatter plot of the network's output versus the desired output. It is also the architecture used by all other algorithms in this chapter for data set B, unless stated otherwise. Each regime's data was then passed through each model in order to measure the cross-prediction error variance, which was then used to calculate the experimental asymmetric KL divergence (4.40), shown in Table 4.2. The distance is smallest between regimes 1 and 2 and largest between regimes 1 and 3. The distances are also more symmetrical than those for data set A.

It is important to note that the KL distances are influenced by the size of the neural networks. Generally, as the size of the networks increases and the models better approximate the regimes, the dis-

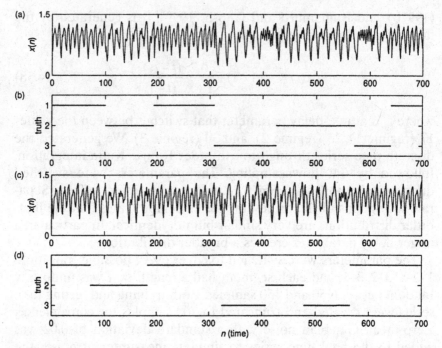

Figure 4.9 Data set B: (a) training, (b) training truth, (c) testing, and (d) testing truth.

tances increase. In general, this is desirable since larger distances imply easier detection of change points. However, one must also be mindful of the bias/variance dilemma. As the size of the networks increase, overfitting can occur in unpredictable ways, reducing the repeatability of the distances. This is especially a problem where some regimes can be modeled with smaller networks than others. We did not use different size networks, however, because such prior knowledge is not available in practice for the unsupervised methods presented later. Also, because neural networks possess many local

Table 4.2 Data Set B: Experimental KL Divergence

Data / Model	1	2	3
1	0	14.7	38.1
2	14.8	0	16.8
3	36.8	16.8	0

minima, we made repeated measurements of the cross-prediction error variances and then used the smallest to calculate the final KL divergences.

It becomes particularly difficult to compare segmentation and identification algorithms using real data. Unlike the prediction task where algorithms can be rigorously compared on the basis of their ability to predict previously unseen test data, there is often no "ground truth" segmentation with which to compare. The same danger exists in employing human expert scoring. For these reasons, we postpone a comparison using real data until a future date.

4.2.10 Simulations

We must first caution that simulations are not intended to prove conclusively that one algorithm is better than another but rather to highlight some of their strengths and weaknesses. There are several reasons for this. First, our implementations are usually based on the simplest formulation of an algorithm. Second, our implementations may deviate slightly from the designer's intent due to insufficient details in the original publication or the use of our own heuristics. Third, particularly in the case of artificial data, we know *a priori* many of the important design parameters such as the number of regimes, embedding dimension, and the like, and thus primarily test the algorithms under optimal conditions. Fourth, we generally display the best of three runs. However, three runs is not sufficient to test convergence.

Using the neural networks trained for the KL measurements, we segmented the test time series using the modified multiple-hypothesis CUSUM algorithm, the results of which are shown in Figs. 4.10 and 4.11. The segmentation is nearly perfect, with no change point error larger than two samples for data set A and three samples for data set B. By comparing the KL distances with the decision functions, one observes that when there is a change between regimes with a large KL distance, detection occurs within a few samples, while between two regimes with a small KL distance there is a visible detection delay.

4.2.11 Why Do the Algorithms Work with Chaotic Data?

It is worthwhile pausing to ask why algorithms that were developed for detecting and classifying changes in piecewise stationary sto-

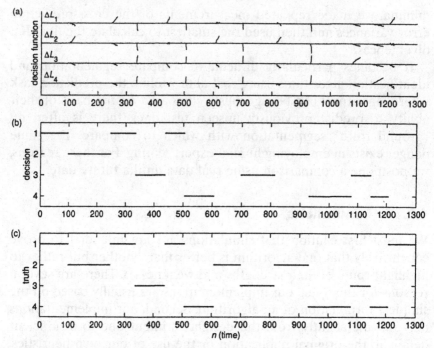

Figure 4.10 CUSUM, data set A, testing: (a) decision function, (b) decision, and (c) truth.

chastic processes also work for switching chaotic signals. In fact, since chaotic signals appear deterministic in the short term and random in the long term, we might have even been tempted to apply the theory for detecting deterministic signals and employ a bank of matched filters.

If we retrace our steps, we see that the most vulnerable assumption made was that the time series are finite-order Markov processes. As previously discussed, a characteristic of such processes is that they have finite memory, such that only a finite number of past samples influence the current value. Thus, if we are predicting the current value starting with a small number of past samples, and increase the number one delay at a time, the prediction improves only to a certain extent and then levels off.

Chaotic signals, on the other hand, effectively possess infinite memory because the effect of past values never dissipates. Nevertheless, for chaotic signals there is a parallel *behavior* to that which we observe as we try to predict finite-order Markov processes, but

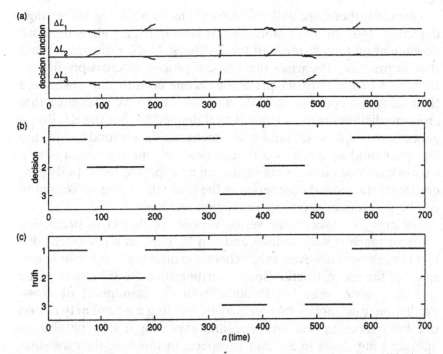

Figure 4.11 CUSUM, data set B, testing: (*a*) decision function, (*b*) decision, and (*c*) truth.

this behavior occurs for a completely different reason. According to Takens' (1980) embedding theorem, if we view the vector of delayed values as a path in a state space, called the lag space, there is a critical dimension, called the embedding dimension, at which point there is a diffeomorphism between the lag space and the original physical state space. At the embedding dimension the trajectories in this so-called reconstruction state space never intersect. Thus, if we are predicting the current value starting with a small number of past samples, and increase the number one delay at a time, the trajectories will intersect less and less, improving the overall prediction, but only to a certain point (the embedding dimension).

Thus, from the standpoint of single-step prediction, stochastic Markov processes and chaotic signals behave similarly as the lag dimension is increased. This is the basis for our statement that chaotic processes are indistinguishable from other nonlinear processes for single-step prediction and is one reason the change detection algorithms work for chaotic processes.

However, there are still other concerns in applying the change detection algorithms to chaotic signals. Recall that the algorithms are dependent on the existence of the Kullback–Leibler divergence and that, in practice, this arises through the process of cross-prediction; training a neural network predictor on one class of time series and then using it to predict another class. When can we guarantee that cross-prediction error variance is well defined so that the KL divergence is a valid distance measure? Stochastic signals tend to "fill" the lag space, and so if the two time series are long enough, a neural network trained on one time series can be expected to properly generalize on the second time series, in the sense that the true likelihood can be determined from the prediction residuals.

For ergodic chaotic time series, we believe that cross-prediction error variance is well defined and can be used as a consistent distance measure. However, other chaotic signals may reside in a subspace of the reconstruction space. Furthermore, the subspaces of the two time series may be disjoint. From the standpoint of cross-prediction, this implies that we may be training a neural network on one time series in one subspace, and expecting it to generalize to another time series in another subspace. In this case the cross-prediction error will be completely unpredictable. Although the KL divergence may not always be well defined for chaotic signals, we can say with some confidence that the cross-prediction error variance will always be smallest when a predictor encounters a time series generated from the same dynamical system on which it was trained. This is in fact why the change detection algorithms work for chaotic signals. We must caution, however, that many of the theoretical properties of these algorithms that depend on KL divergence, such as detection delay, are no longer reliable.

The problem with using cross-prediction as part of a distance measure for chaotic processes can actually be traced back to the presence of the logarithm in the definition of KL divergence, which "blows up" in regions where there is no probability of the signal in the lag space. This problem with KL divergence is well known and has led to several other distance measures between distributions, some of which may be more appropriate for chaotic processes (Diks 1996). We will not further consider alternative metrics in this chapter, however, whenever an algorithm utilizes KL divergence, the possibility of substituting other distance measures should be kept in mind.

4.3   UNSUPERVISED SEQUENTIAL SEGMENTATION

The algorithms of the previous section were based on prior knowledge of the candidate random processes. What if no prior information is available? Is it still possible to segment a single realization of a switching random process? That is, given a single realization of a piecewise ergodic random process, consisting of a finite ordered set of vectors, can we segment the data into contiguous stationary regions? This defines the off-line segmentation problem. We do not know how many stationary regions are contained in the data, nor the change points between regions. Because of this, the problem cannot be easily formulated as a global test between known hypotheses. The only possible approach then is to perform *sequential detection* of change points, which makes the off-line segmentation problem similar to the on-line case, reducing the problem to one of detecting change points between successive pairs of stationary regions.

Basically, there are two closely related approaches to unsupervised sequential segmentation. In the first approach, the detection algorithm is temporarily "turned off" while a model is developed for the incoming data. It is then "turned on" and the incoming data is monitored for a minimum change from the reference model, using some appropriate distance measure.

The second approach, and the one that we will describe in some detail, utilizes the generalized likelihood ratio (GLR). It was first applied to time series by Willsky and Jones (1976), but the approach presented here is more closely associated with Appel and Brandt (1982). It is a Neyman–Pearson test for deciding between two hypotheses: the null hypothesis, H_0, that no change occurred within n samples, and the posited hypothesis, H_1, that a change did occur at the intermediate change point time T. It differs from the standard likelihood ratio test previously described in that the pdf's are unknown and must be estimated directly from the data within their respective regions, as defined by the hypothesized change point time. In addition, a search must be done for the most likely change point within the block of n samples. The algorithm then proceeds analogously as for the supervised case. When the decision function exceeds a preset threshold, a change is detected. The exact transition time is then determined, and the algorithm is reset and started anew.

4.3.1 Testing for a Change in a Memoryless Process

Consider the problem of detecting a change in the first-order pdf of an i.i.d. process. We assume that the data both before and after change can be described by a parametric family of pdf's, $p(x;\theta)$, characterized by the parameter set θ. Since we are interested in an unsupervised algorithm, we do not know the parameters of the pdf before the change point, θ_0, or after the change point, θ_1, nor the change point time, T. The log-likelihood ratio between the hypothesis H_1 that a change occurred at time T and the null hypothesis H_0 that no change occurred within n samples is

$$L(n) = \log\left[\frac{\prod_{m=1}^{T-1} p(x(m);\theta_0')\prod_{m=T}^{n} p(x(m);\theta_1)}{\prod_{m=1}^{n} p(x(m);\theta_0)}\right]$$

$$= \sum_{m=1}^{T-1}\log\left[\frac{p(x(m);\theta_0')}{p(x(m);\theta_0)}\right] + \sum_{m=T}^{n}\log\left[\frac{p(x(m);\theta_1)}{p(x(m);\theta_0)}\right] \quad (4.59)$$

In order to keep the notation simple, the time index here is relative to the last detected change point. This is indicative of the sequential nature of the algorithm, and the fact that all acquired knowledge is discarded following the detection of a change point. Also note that we differentiate between the parameter set under the null hypothesis, θ_0, and the parameter set before change under the change hypothesis, θ_0', since they are estimated over different times.

A decision function is formed from $L(n)$ by replacing the parameter sets θ_0, θ_0', and θ_1, by their maximum-likelihood (ML) estimates, $\hat{\theta}_0$, $\hat{\theta}_0'$, and $\hat{\theta}_1$, over their respective regions, $\{1,\ldots,n\}$, $\{1,\ldots,T-1\}$, and $\{T,\ldots,n\}$, and then maximizing with respect to the change time T:

$$Q(n) = \max_T[L(n)|_{\hat{\theta}_0,\hat{\theta}_0',\hat{\theta}_1}] \quad (4.60)$$

A change is deemed to have occurred within the block of n data if the decision function exceeds a predefined threshold

$$Q(n) \underset{H_0}{\overset{H_1}{\gtrless}} \beta \quad (4.61)$$

Once a change is detected, the change point time can be estimated as the time index that maximized the decision function

$$\hat{T} = \text{argmax}_T[L(n)|_{\hat{\theta}_0,\hat{\theta}_0',\hat{\theta}_1}] \qquad (4.62)$$

The justification for (4.62) is a Bayesian analysis using noninformative priors (see Ruanaidh and Fitzgerald 1996), which demonstrates that the posterior distribution of the change point is closely related to the GLR, $p(T) \propto L(n)|_{\hat{\theta}_0,\hat{\theta}_0',\hat{\theta}_1}$, and that (4.62) is thus equivalent to a *maximum a posteriori* (MAP) estimation. This can be exploited to estimate the uncertainty in a detected change point by, for example, finding the standard deviation of $p(T)$.

The algorithm starts from the last detected change point, from which a small block of n points is examined. To find $Q(n)$, the two hypotheses are compared using (4.59) and (4.60) for all possible change points T within the block, and the maximum value of $L(n)$ is assigned to $Q(n)$. The behavior of $Q(n)$ is such that under H_0, any two blocks of data are equally likely and the decision function will tend to fluctuate near zero, while under H_1 the decision function increases with increasing n. Thus, if $Q(n)$ is less than the threshold, no change point is detected, n is incremented, a new sample is brought into the window, and the process repeats. If $Q(n)$ is greater than the threshold, the algorithm is reset and started anew. A symbolic diagram of this is shown in Fig. 4.12a. Regions where the number of samples is affected by T are shown with solid arrows, while regions affected by n are shown with hollow arrows.

Having defined the algorithm, let us take a closer look at the generalized log-likelihood ratio:

$$L(n)|_{\hat{\theta}_0,\hat{\theta}_0',\hat{\theta}_1} = \sum_{m=1}^{T-1} \log\left[\frac{p(x(m);\hat{\theta}_0')}{p(x(m);\hat{\theta}_0)}\right] + \sum_{m=T}^{n} \log\left[\frac{p(x(m);\hat{\theta}_1)}{p(x(m);\hat{\theta}_0)}\right] \qquad (4.63)$$

In spite of the seeming complexity of (4.63), it often has a fairly simple form in practice because the log-likelihood of a model is evaluated over the same data that is used to determine its parameter set. Using the relative frequency interpretation of expectation, the summations can be approximated by expectations

$$L(n)|_{\hat{\theta}_0,\hat{\theta}_0',\hat{\theta}_1} \approx (T-1)E_{\hat{\theta}_0'}\left[\log\left[\frac{p(x(m);\hat{\theta}_0')}{p(x(m);\hat{\theta}_0)}\right]\right]$$

$$+(n-T+1)E_{\hat{\theta}_1}\left[\log\left[\frac{p(x(m);\hat{\theta}_1)}{p(x(m);\hat{\theta}_0)}\right]\right] \qquad (4.64)$$

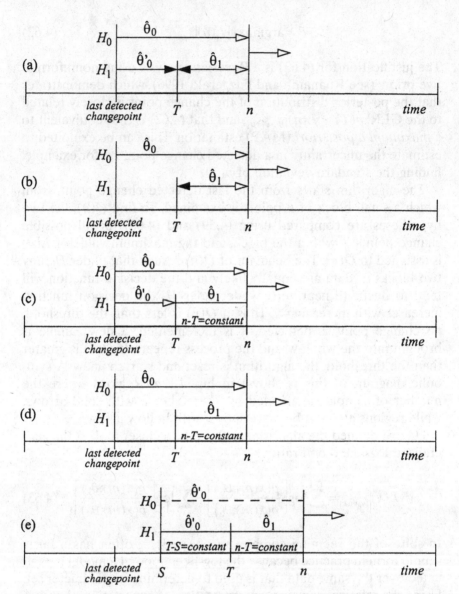

Figure 4.12 GLR change point detection: (*a*) three-model, (*b*) two-model, (*c*) three-model with fixed-length region for θ_1, (*d*) two-model with fixed-length region for θ_1, and (*e*) two-model with fixed-length regions for θ'_0 and θ_1.

Recalling the definition of the asymmetric Kullback–Leibler divergence, we arrive at

$$L(n)|_{\hat{\theta}_0, \hat{\theta}_0', \hat{\theta}_1} \approx (T-1)K_{\hat{\theta}_0', \hat{\theta}_0} + (n-T+1)K_{\hat{\theta}_1 \hat{\theta}_0} \tag{4.65}$$

Thus, the GLR algorithm effectively maximizes the KL divergence between the maximum-likelihood estimates of the pdf representing the entire block of data and those of the subblocks, weighted by the number of samples.

4.3.2 Testing for a Change in a Process with Memory

The hypotheses are exactly the same as in the memoryless case, except now we must evaluate the likelihood recursively. Using Bayes' theorem (4.19), there results

$$L(n) = \sum_{m=1}^{T-1} \log\left[\frac{p_0'(x(m)|X(m-1))}{p_0(x(m)|X(m-1))}\right]$$
$$+ \sum_{m=T}^{n} \log\left[\frac{p_1(x(m)|X(m-1))}{p_0(x(m)|X(m-1))}\right] \tag{4.66}$$

where, for convenience, we have temporarily used a nonparametric representation for the pdf's.

One can question why the second term is conditioned on all prior data even though we are testing for a change point at the intermediate time T. The answer is closely related to the two production models discussed previously. Under the parameter switching model, the system state memory remains intact at the change point, and thus the history of the process before the change point can influence its future and (4.66) is appropriate. Under the external switching model, the history of the process before the change point cannot influence its future. In this case, the second term in (4.66) should be conditioned only on the data since the change point T. In practice, the difference between these two formulations is slight if the processes have a finite effective memory, as previously discussed. In particular, if the maximum effective memory is M, then (4.66) can be written as

$$L(n) \approx \sum_{m=1}^{T-1} \log\left[\frac{p_0'(x(m)|\chi(m-1))}{p_0(x(m)|\chi(m-1))}\right]$$
$$+ \sum_{m=T}^{n} \log\left[\frac{p_1(x(m)|\chi(m-1))}{p_0(x(m)|\chi(m-1))}\right] \tag{4.67}$$

where it is understood that $p(x(1)|\chi(0)) \equiv p(x(1))$. The differences between the two production models now exists only in a region of M samples after the hypothetical change point T.

Assume we can develop predictors for each of the three regions up to a random term

$$x(n) = f(\chi(n-1); \theta_i) + e_i(n) \qquad i = 0, 0', 1 \qquad (4.68)$$

where θ_i now represents the parameter set of a parametric function. The conditional pdf of $x(n)$ can now be written in terms of the pdf of the innovations

$$p_i(x(n)|(\chi(n-1))) = p_{e_i}(e_i(n)) \qquad (4.69)$$

However, it is only when the innovations are truly i.i.d. that we can rewrite (4.66) as

$$L(n) \approx \sum_{m=1}^{T-1} \log\left[\frac{p_{e_0'}(x(n) - f(\chi(n-1); \theta_0'))}{p_{e_0}(x(n) - f(\chi(n-1); \theta_0))} \right]$$
$$+ \sum_{m=T}^{n} \log\left[\frac{p_{e_1}(x(n) - f(\chi(n-1); \theta_1))}{p_{e_0}(x(n) - f(\chi(n-1); \theta_0))} \right] \qquad (4.70)$$

where the approximation ensues due to our ignoring the details of initial conditions, transitions, and possible nonindependence of residuals.

Just as in the memoryless case, the decision function is then formed by replacing the functional parameters, θ_0, θ_0', and θ_1, in (4.68) by their maximum-likelihood (ML) estimates, $\hat{\theta}_0$, $\hat{\theta}_0'$, and $\hat{\theta}_1$, respectively, over their respective regions, $\{1, \ldots, n\}$, $\{1, \ldots, T-1\}$, and $\{T, \ldots, n\}$, and then maximizing with respect to the change point T, exactly as in (4.60). Note that if the prediction errors are modeled as Gaussian, then the ML estimates are equivalent to mean-square-error estimates of the model parameters over their respective regions. Also, the pdf's of the prediction errors may contain "nuisance parameters," such as the covariance, that themselves may have to be replaced by their ML estimates in (4.60); however, they can all be lumped together with the predictor parameters as part of a larger parameter set.

Again, let us take a closer look at the generalized likelihood ratio:

$$L(n)|_{\hat{\theta}_0,\hat{\theta}'_0,\hat{\theta}_1} \approx \sum_{m=1}^{T-1} \log\left[\frac{p_{e'_0}(x(n)-f(\chi(n-1);\hat{\theta}'_0))}{p_{e_0}(x(n)-f(\chi(n-1);\hat{\theta}_0))}\right]$$
$$+ \sum_{m=T}^{n} \log\left[\frac{p_{e_1}(x(n)-f(\chi(n-1);\hat{\theta}_1))}{p_{e_0}(x(n)-f(\chi(n-1);\hat{\theta}_0))}\right] \qquad (4.71)$$

Just as for the memoryless case, this often has a surprisingly simple form due to the fact that the likelihood of a model is evaluated over the same data used to determine its parameters. Using the relative frequency interpretation of summation, (4.71) can be approximated in terms of the KL divergence between two random processes

$$L(n)|_{\hat{\theta}_0,\hat{\theta}'_0,\hat{\theta}_1} \approx (T-1)K_{\hat{\theta}'_0\hat{\theta}_0} + (n-T+1)K_{\hat{\theta}_1\hat{\theta}_0} \qquad (4.72)$$

The generalized likelihood ratios in (4.71) and (4.72) are not equal. However, (4.72) is guaranteed to be nonnegative while (4.71) is not, and in our experience (4.72) is a smoother function of T, although we have no theoretical proof of this.

The expressions for the generalized likelihood ratio in terms of the KL distance in (4.65) and (4.72) can be used as starting points for substituting other distance metrics that may perform better than the KL divergence in some circumstances.

4.3.3 Simplifications

The GLR algorithm is computationally intensive. For each block of n points, on the order of n new ML estimates must be performed. The algorithm thus slows with time if no change point is detected. Extensive use of recursive ML estimates can significantly reduce the computational load but require significant memory. A minor simplification can be made by assuming that the ML estimates of the before-change parameters are approximately equal, $\hat{\theta}'_0 \approx \hat{\theta}_0$, in which case the first terms in (4.59) and (4.70) are then zero. This is called the two-model GLR algorithm. A natural question then arises as to whether to use $\hat{\theta}_0$ or $\hat{\theta}'_0$ in the evaluation of the GLR. Since $\hat{\theta}_0$ is estimated using all the currently available data, it can be estimated more reliably than $\hat{\theta}'_0$ if there is no change point within the current window, resulting in fewer false alarms. In addition, it only needs to be estimated once for each evaluation of the decision function in (4.60). However, once a change point enters the window, the estimate $\hat{\theta}_0$ will start to degrade as a result of trying to model two different regimes, possibly affecting the detection of the change point. Thus, using $\hat{\theta}_0$ will result in fewer false alarms, while using $\hat{\theta}_0'$ will result in more reliable detection. The symbolic diagram shown in Fig. 4.12b shows the case when $\hat{\theta}_0$ is used. For a memoryless process the GLR in (4.59) becomes

$$L(n)|_{\hat{\theta}_0,\hat{\theta}_1} \approx \sum_{m=T}^{n} \log\left[\frac{p(x(m);\hat{\theta}_1)}{p(x(m);\hat{\theta}_0)}\right] \approx (n-T+1)K_{\hat{\theta}_1\hat{\theta}_0} \qquad (4.73)$$

while for the memory case, the GLR in (4.70) becomes

$$L(n)\bigg|_{\hat{\theta}_0,\hat{\theta}_1} \approx \sum_{m=T}^{n} \log\left[\frac{p_{e_1}(x(m)-f(\chi(m-1);\hat{\theta}_1))}{p_{e_0}(x(m)-f(\chi(m-1);\hat{\theta}_0))}\right]$$

$$\approx (n-T+1)K_{\hat{\theta}_1\hat{\theta}_0} \qquad (4.74)$$

Note that the GLR is only evaluated over the last $n - T + 1$ samples.

A more dramatic computational savings can be achieved by eliminating the maximization over T in (4.60) by fixing the length of the data segment associated with the parameter set θ_1, resulting in the symbolic diagram shown in Fig. 4.12c. Eliminating the maximization over T also allows for on-line recursive estimation of all parameters. These two simplifications can be combined, resulting in the symbolic diagram of Fig. 4.12d. The greatest simplification results by using $\hat{\theta}_0'$ in place of $\hat{\theta}_0$ in (4.73) or (4.74), and using a fixed-length estimation region for $\hat{\theta}_0'$, in effect monitoring the distance between the data contained within two adjoining fixed-length windows. This is shown in Fig. 4.12e. The windows are usually chosen to be the same length.

Once a change point is detected it is very important to properly estimate the change time using the full three-model GLR, searching in a limited region around the initial detection time. Any error in estimating a change point degrades the model estimates for the beginning of the next detection stage, and this can propagate all the way through the data set. Missing a change point altogether can have serious consequences for the algorithm because the null model will then be simultaneously trying to fit two different regimes. The exception to this is the two fixed windows approach of Fig. 4.12e, where the consequences of a missed detection disappear once the change point moves outside the scope of the two windows.

4.3.4 Examples

As an example, consider the problem of detecting a change in a memoryless multivariate Gaussian process $x(n) \sim N(\mu, \Sigma)$. In this case, it turns out that the generalized log-likelihood ratio from either (4.63) or (4.65) are the same

$$2L(n)|_{\hat{\theta}_0, \hat{\theta}_0, \hat{\theta}_1} = (T-1)\log\left[\frac{|\Sigma_0|}{|\Sigma_0'|}\right] + (n-T+1)\log\left[\frac{|\Sigma_0|}{|\Sigma_1|}\right] \quad (4.75)$$

where, for example, Σ_0 is the ML estimate of the covariance over the region $\{1, \ldots, n\}$

$$\Sigma_0 = \frac{1}{N}\sum_{m=1}^{n}[x(m) - \mu_0][x(m) - \mu_0]^{\dagger} \qquad \mu_0 = \frac{1}{N}\sum_{m=1}^{n}x(m) \quad (4.76)$$

and the other covariances are similarly estimated over their respective regions. Equation (4.75) indicates that a change in the mean and a change in the covariance are both detected through a change in the covariance. This is intuitive because a pure change in the mean creates an increase in the covariance estimate for the whole region $\{1, \ldots, n\}$. However, this also implies that this test cannot differentiate between a pure change in the mean and a pure change in the covariance, which may be undesirable in some situations. We shall have more to say about this shortly. For the two-model case, we use (4.73) and substitute the appropriate KL divergence from (4.37):

$$2L(n)|_{\hat{\theta}_0, \hat{\theta}_1} \approx (n-T+1)\Big\{[\mu_1 - \mu_0]^{\dagger}\Sigma_0^{-1}[\mu_1 - \mu_0]$$

$$+ \mathrm{Tr}[\Sigma_1\Sigma_0^{-1}] - D + \log\left[\frac{|\Sigma_0|}{|\Sigma_1|}\right]\Big\} \quad (4.77)$$

It is important to recognize that this is not the same as setting $\Sigma_0' = \Sigma_0$ in (4.75).

As a second example, consider testing for a change in a scalar process with memory using a parametric predictive model under the assumption of Gaussian innovations with variance σ^2. The most important implication of this assumption is that maximum likelihood is now equivalent to mean square error, and we can use a standard MSE criteria for training the models. The generalized log-likelihood ratio from (4.71) is easily shown to be

$$2L(n)|_{\hat{\theta}_0,\hat{\theta}'_0,\hat{\theta}_1} \approx (T-1)\log\left[\frac{\sigma_0^2}{\sigma'^2_0}\right] + (n-T+1)\log\left[\frac{\sigma_0^2}{\sigma_1^2}\right] \quad (4.78)$$

This equation can also be used with nonlinear models where no explicit representation of the KL divergence is available. The two-model generalized likelihood ratio for Gaussian residuals follows from (4.74):

$$2L(n)|_{\hat{\theta}_0,\hat{\theta}_1} \approx (n-T+1)\left\{\frac{1}{\sigma_0^2}\frac{1}{(n-T+1)}\sum_{m=T}^{n}(x(m)\right.$$
$$\left. -f(\chi(m-1);\theta_0))^2 - \log\left[\frac{\sigma_1^2}{\sigma_0^2}\right] - 1\right\} \quad (4.79)$$

Note that the summation is the cross-prediction error variance of the data from the second window measured against a model built using all available data. Again, it is important to recognize that this is not the same as setting $\sigma'_0 = \sigma_0$ in (4.78).

As a subset of the above case, consider testing for a change in a Gaussian autoregressive (AR) process. Recall that if a process is modeled as purely AR, then the optimal predictor is a finite impulse response (FIR) filter. The generalized log-likelihood ratio in the two-model case is, from (4.74) and (4.42),

$$2L(n)|_{\hat{\theta}_0,\hat{\theta}_1} \approx (n-T+1)\left\{\frac{1}{\sigma_0^2}(\mathbf{w}_1-\mathbf{w}_0)^\dagger\mathbf{R}_1(\mathbf{w}_1-\mathbf{w}_0)\right.$$
$$\left. +\frac{\sigma_1^2}{\sigma_0^2} - \log\left[\frac{\sigma_1^2}{\sigma_0^2}\right] - 1\right\} \quad (4.80)$$

where $\mathbf{R}$ is the Toeplitz autocorrelation matrix and $\mathbf{w}$ is a vector of optimal linear prediction coefficients.

All these tests are sensitive to changes in any of the model parameters. In some cases, it may be desirable to create separate tests for each of the parameters or particular subsets. For example, it may be desirable to create a test that is insensitive to pure changes in signal power, for the purpose of detecting pure changes in the mean of an i.i.d process or the spectrum of an AR process. This can be accomplished by treating the parameter to be isolated as a constant during the calculation of the generalized log-likelihood ratio. In some cases, additional assumptions about the isolated parameters may be required so that they completely decouple from the other parameters. Using the two AR model case as an example, assuming constant excitation power results in

$$L(n)|_{\hat{\theta}_0,\hat{\theta}_1} \propto (n - T + 1)(\mathbf{w}_1 - \mathbf{w}_0)^\dagger \mathbf{R}_1(\mathbf{w}_1 - \mathbf{w}_0) \qquad (4.81)$$

Note that this is the same as setting $\sigma_0 = \sigma_1$ in (4.80). In practice, the test in (4.81) merely reduces the sensitivity to power changes but does not completely eliminate it.

For nonlinear models, a pure change in signal power may not result in a proportional change in prediction error variance. This makes it more difficult to design decision functions that are insensitive to pure changes in signal power.

4.3.5 Discussion

The GLR algorithm is very powerful and its primary strength is that it makes few assumptions about the data, other than that it should be piecewise ergodic. Although theoretical analysis is difficult, Deshayes and Picard (1986) have investigated its asymptotic properties for certain families of distributions. Its primary limitation is in detecting rapid switching between regimes. In practice, the maximization over the change point T in (4.60) must be restricted in order to avoid unreliable estimates for small samples. This creates a dead zone upon restarting the algorithm after detection of the last change point. The greater the complexity of the parametric model, the larger the required dead zone. For detecting a change in the mean, the dead zone can be relatively short, on the order of 5 or 10 samples, depending on the acceptable false alarm rate. However, for detecting a change in a process with memory, such as an AR process, the dead zone must be much larger in order to reliably estimate correlations. The situation is even worse when the use of nonlinear predictors are required.

While the GLR is directly applicable to the use of nonlinear predictors, such as neural networks, it is computationally intensive. Even for the two-model GLR, a single nonlinear model must be trained approximately n (minus the dead zone) separate times just to evaluate the decision function for a block of n samples. In addition, neural networks cannot be trained in a single pass through a block of data and possess many local minima, necessitating a comparison among multiple runs. For these reasons, the training time can become prohibitive. This implies that for the initial detection of a change point, only the fixed block length versions of Fig. 4.12c, Fig. 4.12d, or Fig. 4.12e are feasible. However, as stated previously, the full three-model GLR should still be used to estimate the exact change point within a limited region around the initial detection point.

A related approach (Schreiber 1997) is to divide the entire time series into equal-length segments, train separate nonlinear predictors on each segment, and then create a matrix of cross-prediction error variances by passing the data for each segment through each of the trained models. However, any given segment may contain more than one regime, distorting the distance measurements. Still, the method is much faster than the GLR and can be used to help choose some of the parameters of the GLR. However, rather than the raw cross-prediction error variances, the experimental KL divergence interpretation of (4.40) should be used.

Another drawback to the GLR is that, being a sequential detector, it discards information from previous stationary regions. If we suspect *a priori* that the number of potential regimes is limited and that some regimes may be revisited, we can potentially improve the estimate of our parametric models and reduce the dead zone by using global (in time) information. This is clearly a noncausal approach but is not a serious drawback since we are interested in the off-line segmentation problem. Starting with the next section we consider algorithms that use the entire data set for model building.

4.3.6 Simulations

For both data sets, we used the two-model GLR with fixed-length regions, as in Fig. 4.12e, and the *symmetric* KL distance as the decision function. For data set A the fixed length was 30 samples, while for data set B it was 40 samples. The window was expanded by 5 samples if no detection occurred. Once a detection occurred, we did coarse and then fine searches near the detection point using the full three-model GLR (4.78). The training data was only used to set the threshold, and then the algorithm was run on the test set with no other adjustments.

Starting with data set A, Fig. 4.13a shows the decision function and Fig. 4.13b the search for the optimal changepoint (4.78). The times when there is no decision function represents the dead zone after a detection. From Fig. 4.13b we see that the estimates of the change point times are very accurate, with no change point error larger than

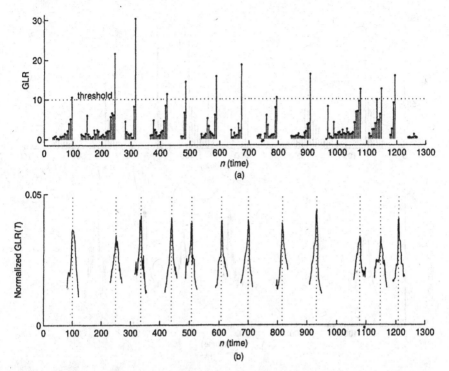

Figure 4.13 GLR, data set A, testing: (*a*) decision function and (*b*) change point search (solid) and true change point (vertical dotted).

two time steps. Furthermore, the GLR as a function of time around the change points is highly peaked, indicating a high degree of confidence in the estimated change points. Note, however, that the GLR provides no information about recurring regimes.

For data set B, Fig. 4.14*a* shows the decision function and Fig. 4.14*b* the search for the optimal change point (4.78). The results are good except for the missed detection of a change at sample 182.

4.4 MEMORYLESS MIXTURE MODELS

We now seek an unsupervised segmentation algorithm that overcomes some of the drawbacks of the GLR algorithm. As a starting point, we assume that the data are generated by switching among a *finite* number of regimes. Some of the regimes may be active at discontiguous times (see Fig. 4.1), and therefore the algorithm should make use of the *entire* time series for improved model estimation

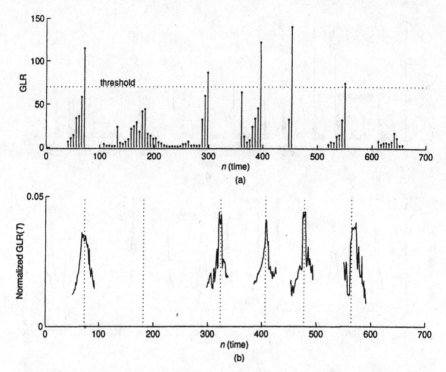

Figure 4.14 GLR, data set B, testing: (*a*) decision function and (*b*) change point search (solid) and true change point (vertical dotted).

prior to or in conjunction with the segmentation stage. In addition, some of the regimes may involve nonlinear processes, and therefore the algorithm must be computationally tractable for parametric nonlinear models that cannot be trained in a single step. Taking all these constraints into consideration, we propose to first train a global mixture model for the entire random process and then use the global information to deduce information about local change points. The local change point information can then be used in turn to refine the global models. Although the scope of this book is nonlinear dynamical systems, we first consider mixture models appropriate to memoryless random processes to highlight the basic ideas and then consider processes with memory in the next section.

4.4.1 Memoryless Models

Let us consider modeling a vector i.i.d. process that may be multimodal. That is, at each time instant a vector $\mathbf{x}(n)$ may be generated

by one of K different distributions. Let $p(x, k)$ be the joint probability of a vector $\mathbf{x}$ belonging to the discrete mode k. We cannot
determine this distribution directly from the data because we do not
know which modes are active at which times. We can, however,
observe the marginal density

$$p(\mathbf{x}) = \sum_{k=1}^{K} p(\mathbf{x},k) = \sum_{k=1}^{K} P(k)p(\mathbf{x}|k) \equiv \sum_{k=1}^{K} P(k)p_k(\mathbf{x}) \qquad (4.82)$$

where the kth mixing coefficient, $P(k)$, is the *a priori* probability of
the kth mode, and we have used our earlier notation $p_k(\mathbf{x}) \equiv p(\mathbf{x}|k)$.
Since it is known that a sum of Gaussian distributions can approxi-
mate any continuous distribution, the modes are usually modeled as
Gaussian:

$$p_k(\mathbf{x}) = \frac{1}{(2\pi)^{D/2}|\Sigma_k|^{1/2}} \exp\left[-\frac{1}{2}(\mathbf{x} - \boldsymbol{\mu}_k)^\dagger \Sigma_k^{-1}(\mathbf{x} - \boldsymbol{\mu}_k)\right] \qquad (4.83)$$

where D is the dimension of $\mathbf{x}$.

It is worthwhile pausing to consider some *production models* that
are compatible with the mixture in (4.82):

- Choose a regime k^* according to the prior probabilities
  $P(k)$ and then generate a single sample from the distribution
  $p_{k^*}(\mathbf{x})$.
- Choose a regime k^* according to the prior probabilities $P(k)$
  and then generate N independent samples from the distribution
  $p_{k^*}(\mathbf{x})$.
- Choose a regime k^* with uniform probability and then gener-
  ate $N \cdot P(k^*)$ independent samples from the distribution $p_{k^*}(\mathbf{x})$.

Each of these three production models is consistent with the mixture
model in (4.82), and yet the first one results in a stationary process,
while the latter two result in piecewise stationary processes. In fact,
it is this very ambiguity that allows us to separate the modeling from
the segmentation. Thus, fitting a time series to the model represented
by (4.82) will not by itself give us any segmentation information.
Later on we will show how we can obtain segmentation information
from the fitted model parameters. However, we must first discuss
how to determine the model parameters, given a time series $x(n)$,
$n = 1, \ldots, N$. One approach is to maximize the likelihood

$$L = \prod_{n=1}^{N} p(\mathbf{x}(n)) = \prod_{n=1}^{N} \sum_{k=1}^{K} g_k p_k(\mathbf{x}) \qquad (4.84)$$

with respect to the free parameters, which consists of the parameters of the K distributions and the mixing coefficients, which in the context of estimation we label as $g_k \equiv P(k)$. Unfortunately, this is difficult to maximize using standard techniques such as gradient descent. The expectation–maximization (EM) algorithm (Dempster et al. 1977), on the other hand, is ideally suited to estimating joint distributions when some of the variables are not directly observable. The EM algorithm postulates a "complete" joint distribution of the observed and unobserved variables. Given an initial set of model parameters, the expectation step of the algorithm involves finding the distribution of the unobserved variables and then using this distribution to find the expected value of the complete negative log-likelihood, which then becomes the cost function. The maximization step involves minimizing the resulting cost function with respect to the model parameters. The algorithm then repeats until convergence. Since it can be shown that the cost function is decreased at each iteration, convergence to a local minimum is guaranteed. We now present an intuitive development due to Weigend et al. (1995) that highlights the outcomes of the full application of the EM algorithm.

If we knew to which mode each sample belonged, the problem would easily decouple into K separate maximum-likelihood problems. Therefore, we postulate a latent binary variable, $I_k(n)$, which indicates which mode is active at each time step. This allows the "complete" likelihood to be written as

$$L = \prod_{n=1}^{N} \prod_{k=1}^{K} [g_k p_k(\mathbf{x})]^{I_k(n)} \qquad (4.85)$$

Unfortunately, we do not know which mode is active at which times. However, we can estimate this from the current values of the mixture model using Bayes' theorem, in what is known as the expectation (E) step

$$h_k(n) \equiv E[I_k] = P(k|\mathbf{x}(n)) = \frac{P(k)p(\mathbf{x}(n)|k)}{p(\mathbf{x}(n))} = \frac{g_k p_k(\mathbf{x})}{\sum_{j=1}^{K} g_j p_j(\mathbf{x})} \qquad (4.86)$$

The posterior probabilities, h_k, are then used in place of the indicator variables in (4.85), the negative log of which becomes our cost function. For the special case of a Gaussian mixture, the cost function is

$$J = \sum_{n=1}^{N}\sum_{k=1}^{K}\left\{-h_k(n)\log[g_k]+\frac{1}{2}h_k(n)\right.$$

$$\left.\left\{[\mathbf{x}(n)-\boldsymbol{\mu}_k]^\dagger\Sigma_k^{-1}[\mathbf{x}(n)-\boldsymbol{\mu}_k]+\log[|\Sigma_k|]\right\}\right\}+\Gamma\left(1-\sum_{k=1}^{K}g_k\right) \quad (4.87)$$

where we have appended a constraint, using the Lagrange multiplier Γ, that the mixing coefficients must sum to one. Minimization of this cost function with respect to the free parameters, including Γ, forms the maximization (M) step:

$$g_k = \frac{1}{N}\sum_{n=1}^{N}h_k(n) \quad (4.88)$$

$$\boldsymbol{\mu}_k = \frac{1}{\displaystyle\sum_{n=1}^{N}h_k(n)}\sum_{n=1}^{N}h_k(n)\mathbf{x}(n) \quad (4.89)$$

$$\Sigma_k = \frac{1}{\displaystyle\sum_{n=1}^{N}h_k(n)}\sum_{n=1}^{N}h_k(n)[\mathbf{x}(n)-\boldsymbol{\mu}_k][\mathbf{x}(n)-\boldsymbol{\mu}_k]^\dagger \quad (4.90)$$

The update order is important in that the new means in (4.89) are used to calculate the new covariance in (4.90). The algorithm then returns to the E step with the new parameters and is iterated until the parameters converge. Note that using the posterior probabilities as an estimate of the indicator variables has completely decoupled the problem. The means and covariances of the modes are estimated independently of each other and the mixing coefficients are estimated independent of the modes and covariances.

4.4.2 Training Issues

While the EM algorithm is guaranteed to reduce the cost function in (4.87) with each iteration, it does not guarantee finding the global minimum. In fact, even for the simplest of problems there may be many local minima. For this reason, annealing the model parameters during training can greatly improve the repeatability of the results, although not necessarily increase the chances of finding the global minimum. Annealing can take two forms. First, the complexity of the model can be annealed. For example, the covariances can be constrained to be diagonal or even equal to each other early in the training, and then relaxed to maximum freedom. Second, parameters that reflect model confidence can be annealed. For example, early in the training the mixture coefficients can be restricted to reflect equal prior probabilities, or the covariances can be kept artificially large to reflect initial model uncertainty, and then relaxed to full freedom. Such techniques are an *ad hoc* way of placing priors on the model parameters, which is formally accomplished with maximum *a posteriori* (MAP) methods, at the expense of increased algorithmic complexity.

A much more difficult question is how to choose the number of modes. There are basically two approaches. One can start from a small order K, train the system to completion, and then start over and repeat the training at the next model order $K + 1$. Training stops when the likelihood reaches a plateau with respect to model order. The other approach involves annealing the model order during training, using techniques such as tree growing and pruning algorithms. There are numerous examples of this in the literature but they are beyond the scope of this chapter.

Should a model fail to win a sufficient number of samples, the estimate of the covariance in particular can become very unreliable, and for this reason provision should be made to allow a model to gracefully exit the competition if its prior probabilities drop below a predefined threshold.

4.4.3 Segmenting with the Trained Mixture Model

Note that the mixture model is memoryless. That is, one could randomly reorder the time series and the algorithm would converge to the same final parameters, given the same initial conditions. Clearly, there is no reason to expect a global mixture model to correctly recognize that a given sequence is piecewise stationary, unless there is absolutely no overlap of the clusters in the input space. How can we use the mixture model *after training* to segment a time series (either the one it was trained on or a new one)? Assuming we have trained a mixture model with the entire data set, we now have a set of candidate models for the data and can proceed as for the supervised sequential detection and identification presented in Section 4.2. The process log-likelihood ratio between modes i and j up to time n is

$$L_{ij}(n) = \log\left[\frac{\prod_{m=1}^{n} p(\mathbf{x}(m),i)}{\prod_{m=1}^{n} p(\mathbf{x}(m),j)}\right] = \sum_{m=1}^{n} \log\left[\frac{g_i p_i(\mathbf{x})}{g_j p_j(\mathbf{x})}\right]$$

$$= \sum_{m=1}^{n} \log\left[\frac{h_i(m)}{h_j(m)}\right] \tag{4.91}$$

This can be written in a recursive manner, $l_{ij}(n) = l_{ij}(n-1) + l_{ij}(n)$, exactly as in (4.22), except here the log-likelihood increment, $l_{ij}(n)$, is expressed in terms of the posterior probabilities

$$l_{ij}(n) = \log\left[\frac{h_i(n)}{h_j(n)}\right] \tag{4.92}$$

Now that we have a way to calculate the log-likelihood increment, we proceed exactly as for the supervised case; we first identify which mode is "active" using the SPRT and then perform change detection and identification on the entire time series using the CUSUM test. The only adjustment to the change detection tests presented in Section 4.2 is that here the log-likelihood increment is based on the ratio of the posterior probabilities rather than the pdf's themselves. The relationship between the EM algorithm and the CUSUM test is analogous to that between a hidden Markov model (HMM) and the Viterbi algorithm. The HMM provides the models used to evaluate the probabilities associated with a sequence, which are then analyzed by the Viterbi algorithm. Once the time series has been segmented and identified, we can make a final pass through the data to improve the models. We can also attempt to segment previously unseen i.i.d. time series.

Let us take a closer look at the effect of the priors. The log-likelihood increment between the true mode i^* and any other mode j can be decomposed as

$$l_{i^*j}(n) = \log\left[\frac{h_{i^*}(n)}{h_j(n)}\right] = \log\left[\frac{p_{i^*}(\mathbf{x})}{p_j(\mathbf{x})}\right] + \log\left[\frac{g_{i^*}}{g_j}\right] \tag{4.93}$$

The expected value of this with respect to the true mode will be the slope of $L_{i^*j}(n)$

$$E_{i*}[l_{i*j}(n)] = K_{i*j} + \log\left[\frac{g_{i*}}{g_j}\right] \tag{4.94}$$

where K_{i*j}, a nonnegative quantity, is the Kullback–Leibler divegence (4.36) between the conditional pdf's for modes $i*$ and j. The priors can have several effects on the detectability of a mode. In order to guarantee detectability for the multiple-hypothesis CUSUM test, the slope of $l_{i*j}(n)$ must be positive or $\log[g_{i*}/g_j] > -K(p_{i*}, p_j) \forall j$. Since this is difficult to guarantee, it is better to use the modified log-likelihood in (4.6) and (4.16), which is less sensitive to the individual ratios. As long as detectability is guaranteed, then the effect of the priors will be either to increase or decrease the detection delay. Alternatively, one can simply ignore the priors in (4.93) in order to guarantee that every model has the potential for becoming active.

4.5 MIXTURE MODELS FOR PROCESSES WITH MEMORY

We now turn to modeling a process with memory using a mixture model. We will assume a one-dimensional time series, but the results can easily be generalized to the multidimensional case. Recall that the multivariate pdf of a random process, with an effective memory depth of M, can be decomposed as

$$p(X(n)) = \prod_{n=1}^{N} p(x(n)|\chi(n-1)) \tag{4.95}$$

As before, let us entertain the possibility that the random process is produced by K switching regimes. Therefore, we propose the joint conditional density $p(k,x(n)|\chi(n-1))$, where the discrete variable $k = 1, \ldots, K$ indicates the regime. Once again, we cannot observe this pdf directly because we do not know which regime is active at which times, but we can observe its marginal distribution

$$p(x(n)|\chi(n-1)) = \sum_{k=1}^{K} p(k,x(n)|\chi(n-1))$$

$$= \sum_{k=1}^{K} P(k|\chi(n-1))p(x(n)|\chi(n-1),k) \tag{4.96}$$

where $P(k|\chi(n-1))$ is the *a priori* probability of the kth regime, and we have used the more compact notation $p_k(x(n)|\chi(n-1)) \equiv$

$p(x(n)|\chi(n-1), k)$. Note that we have implicitly assumed that each regime is characterized by the same memory depth. This certainly need not be the case. Theoretically, however, we can always choose the overall system memory depth to be the *maximum* memory depth of the individual regimes. Certain regimes will then have more information than they need to predict the next sample. We shall have more to say about this later.

If we can find predictors for each of the subprocesses, up to a random term e_k,

$$x(n) = f(\chi(n-1); \theta_k) + e_k(n) \qquad k = 1, \ldots, K \qquad (4.97)$$

then we can evaluate the conditional pdf in terms of the innovations

$$p_k(x(n)|\chi(n-1)) = p_{e_k}(e_k(n)) \qquad (4.98)$$

The innovations are usually modeled as a Gaussian distribution, which for a one-dimensional time series becomes

$$p_{e_k}(e_k) = \frac{1}{\sqrt{2\pi\sigma_k^2}} \exp\left[-\frac{[x(n) - \hat{x}_k(n)]^2}{2\sigma_k^2} \right] \qquad (4.99)$$

where we have defined $\hat{x}_k(n) \equiv f(\chi(n-1); \theta_k)$ to be kth predictor's estimate of the next value of the time series and σ_k^2, a "nuisance parameter", is the variance of the error of the kth predictor.

Taking the expected value of both sides of (4.96) gives the best MMSE prediction of the next value of the time series

$$\hat{x}(n) \equiv E[x(n)|\chi(n-1)] = \sum_{k=1}^{K} P(k|\chi(n-1))E[x(n)|\chi(n-1), k]$$

$$= \sum_{k=1}^{K} P(k|\chi(n-1))\hat{x}_k(n) \qquad (4.100)$$

which is a weighted sum of outputs of the individual predictors. This can be regarded as the total system output.

It is worthwhile pausing to consider some production models that are compatible with the mixture in (4.96):

• Choose a regime k^* according to the prior probabilities

$$P(k|\chi(n-1))$$

and then generate a single sample from the distribution $p_{k^*}(x(n)|\chi(n-1))$.

- Choose a regime $k*$ according to the prior probabilities $P(k|\chi(n-1))$ and then calculate the next sample as $x(n) = f(\chi(n-1); \theta_{k*}) + e_{k*}(n)$.

Unlike the memoryless case, we cannot generate more than one sample at a time because these production models are recursive due to the dependence of the mixing coefficients, $P(k|\chi(n-1))$, on the recent past of the time series itself. When might the regime be predictable from the time series itself? The most obvious case is when the regime is a *direct function* of the recent past of the time series. The simplest example of this is the threshold autoregressive (TAR) model (Tong and Lim 1980). Alternatively, the regime may be a probabilistic function of the recent past of the time series, such as for the mixAR model (Zeevi et al. 1999), where the mixing coefficients are a function of the recent past of the time series through the multinomial logistic function

$$P(k|\chi(n-1)) = \frac{\exp[\mathbf{s}_k^\dagger \chi(n-1)]}{\sum\limits_{k=1}^{K} \exp[\mathbf{s}_k^\dagger \chi(n-1)]} \qquad (4.101)$$

where $\mathbf{s}_k$ is a vector of weights. A regime $k*$ is chosen according to (4.101), and then the next value of the time series is generated according to one of K autoregressive models

$$x(n) = \mathbf{a}_{k*}^\dagger \chi(n-1) + e_{k*}(n) \qquad (4.102)$$

where $\mathbf{a}_k$ is a vector of AR coefficients. Zeevi et al. (1999) have shown that if the individual AR models have all their poles inside the unit circle, the resulting time series is *stationary*. This raises an important question: If the regime is predictable from the recent past of the time series itself, does this preclude the time series from being considered piecewise stationary? For nonergodic nonlinear dynamical systems, the answer is no. Imagine several regimes occupying nonoverlapping regions in state space. If each regime has the tendency to keep the state within its respective region, then the resulting time series can be considered piecewise stationary. However, this need not be the case, which suggests that fitting a time series to the model represented by (4.96) will not by itself provide any segmentation information, and further analysis is required to determine if the nonlinear dynamical regimes have this property.

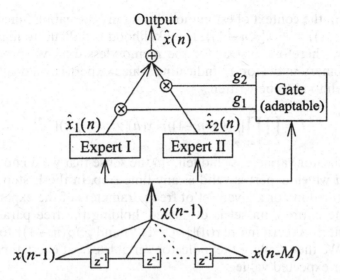

Figure 4.15 Mixture of experts during activation for a network with two experts.

4.5.1 Mixture of Experts

In the mixture of experts (MOE) algorithm[3] (Jordan and Jacobs 1994), the mapping of the prior probabilities is provided by an adaptable function called the *gate*. The particular variation of the algorithm we will examine is called the *nonlinear gated experts* (Weigend et al. 1995), which implements the gate using a single multilayer perceptron, as shown in Fig. 4.15. Note that the individual predictors, called *experts* in this context, and the gate all see the same input.

We now turn to the question of how to train the gate and experts. In the following development, for ease of reading, we leave out explicit time dependence whenever possible. Also, there is an implicit iteration index in all the following equations. Given a time series of length N, we choose the free parameters of the predictors and gate that maximize the process log-likelihood. If the innovations are i.i.d, we can rewrite the process likelihood as

$$p(\mathrm{X}(n)) = \prod_{n=1}^{N} \sum_{k=1}^{K} g_k(\chi(n-1)) p_{e_k}(x(n) - \hat{x}_k(n)) \qquad (4.103)$$

---

[3] The mixture of experts was originally developed in the context of regression and was first applied to direct time-series analysis by Waterhouse and Robinson (1995) and Weigend et al. (1995).

where, in the context of estimation, we signify the gating function by $g_k(\chi(n-1)) \equiv P(k|\chi(n-1))$. The likelihood is difficult to maximize directly. Therefore, just as for the memoryless case, we propose a latent binary indicator, I_k, indicating which expert is valid, allowing the likelihood to be written

$$L = \prod_{n=1}^{N}\prod_{k=1}^{K}[g_k(\chi(n-1))p(x(n)|\chi(n-1),k)]^{I_k(n)} \qquad (4.104)$$

The indicator variable is "hidden," in the sense that we do not know *a priori* which expert is valid at any time step. In the E step of the EM algorithm, for a given set of free parameters of the experts and gate, the entire data set is evaluated, holding the free parameters constant, to determine $p(x(n)|\chi(n-1),k)$ and $g_k(\chi(n-1))$ for all k and n. We then replace the indicator variables, I_k, at every time step, by their expected value:

$$
\begin{aligned}
h_k(n) &\equiv E[I_k(n)] = P(k|x(n),\chi(n-1)) \\
&= \frac{P(k|\chi(n-1))p(x(n)|\chi(n-1),k)}{p(x(n)|\chi(n-1))} \\
&= \frac{g_k(\chi(n-1))p(x(n)|\chi(n-1),k)}{\displaystyle\sum_{k=1}^{K} g_k(\chi(n-1))p(x(n)|\chi(n-1),k)}
\end{aligned}
\qquad (4.105)
$$

Thus, h_k is the posterior probability of expert k, given both the current value of the time series and the recent past. For the M step, (4.104) is maximized or equivalently, the negative log-likelihood,

$$
\begin{aligned}
J = \sum_{n=1}^{N}\sum_{k=1}^{K}\Bigg\{ &-h_k(n)\log[g_k(\chi(n-1))] \\
&+ \frac{1}{2}h_k(n)\left[\frac{[x(n)-\hat{x}_k(n)]^2}{\sigma_k^2} + \log[\sigma_k^2]\right]\Bigg\}
\end{aligned}
\qquad (4.106)
$$

is globally minimized over the free parameters. The process is then repeated. If, in the M step (4.106), is only decreased and not minimized, then the process is called the generalized EM (GEM) algorithm. This is necessary when either the experts or gate are nonlinear, and a search for the global minimum is impractical.

The first term in the summation of (4.106) is closely related to the Kullback–Leibler divergence between the posterior probabilities and the gate. It has a minimum when only one expert is valid and thus encourages the experts to divide up the input space. In order to ensure that the outputs of the gate sum to unity, the output layer of the multilayer perceptron is a multinomial logistic function, often called a *softmax* in a neural network context,

$$g_k(\chi) = \frac{\exp[s_k(\chi)]}{\displaystyle\sum_{j=1}^{K} \exp[s_j(\chi)]} \qquad (4.107)$$

where s_k is the kth input to the softmax. For a gate implemented as a multilayer perceptron, the KL divergence term in (4.106) cannot be minimized in a single step, and the GEM algorithm must be employed. If the gate is trained through gradient descent, the back-propagated error to the *input* side of the softmax, from (4.106) and (4.107), at each time step is

$$\frac{\partial}{\partial s_k} J(n) = g_k(\chi) - h_k \qquad (4.108)$$

This is the same back-propagated error that would result for a mean square error with the posterior probabilities acting as the desired signal. Thus, the posterior probabilities act as targets for the gate. For each EM iteration several training iterations may be required for the gate, since it is implemented using a multilayer perceptron.

There is an analytical solution for the experts at each iteration when they are linear predictors, $\hat{x}_k(n) = \mathbf{w}_k^t \chi(n-1)$,

$$\mathbf{w}_k = \mathbf{R}_k^{-1} \mathbf{p}_k \qquad (4.109)$$

where $\mathbf{R}_k$ and $\mathbf{p}_k$ are weighted autocorrelation and cross-correlation matrices, respectively:

$$\mathbf{R}_k = \frac{1}{\displaystyle\sum_{n=1}^{N} h_k(n)} \sum_{n=1}^{N} h_k(n)\chi(n-1)\chi^t(n-1)$$

$$\mathbf{p}_k = \frac{1}{\displaystyle\sum_{n=1}^{N} h_k(n)} \sum_{n=1}^{N} h_k(n)\chi(n-1)x(n) \qquad (4.110)$$

Note that the solution represents a weighted discrete Wiener (1960) filter, where the data are weighted by the posterior probabilities. If the experts are nonlinear, the gradient of the cost function with respect to the ith weight in the kth expert w_{ki} at each time step is

$$\frac{\partial}{\partial w_{ki}} J(n) = -\frac{h_k(n)}{\sigma_k^2}[x(n) - \hat{x}_k(n)]\frac{\partial}{\partial w_{ki}}\hat{x}_k(n) \qquad (4.111)$$

which has the simple interpretation of weighting the relative error by the posterior probabilities. However, Zeevi et al. (1997) have shown that the nonlinearity of the gate allows the mixture of experts to be a universal approximator, even if the expert predictors are linear. No matter whether the experts are linear or nonlinear, there is an exact solution for the variance at the end of each epoch:

$$\sigma_k^2 = \frac{1}{\sum\limits_{n=1}^{N} h_k(n)} \sum_{n=1}^{N} h_k(n)[x(n) - \hat{x}_k(n)]^2 \qquad (4.112)$$

Note that the update order is important. For a given iteration of the EM algorithm, the variance is updated at the end of the M step *after* the experts have been trained.

4.5.2 Training Issues

For a given data set, there are many local minima on the performance surface. Heuristics that were given for training memoryless mixture models also hold here. Annealing the experts' model complexity, however, is clearly a more difficult task if the experts are nonlinear. In order not to dramatically increase training time, the complexity should be increased without completely discarding the training results of the previous iteration. Weigend et al. (1995) anneals the posterior probabilities that are used to update the variance parameters and found that it led to better repeatability.

As with the number of clusters in the memoryless mixture model, the number of experts can be grown or pruned during training. The primary references in the literature are Fritsch et al. (1997), Jacobs et al. (1997), and Ramamurti and Gosh (1999); the latter have addressed it specifically as it applies to a mixture of expert predictors.

4.5.3 Segmenting with the Trained Mixture of Experts

Let us now return to the application of the mixture of experts to time-series segmentation. Recall that for the memoryless case, the mixture model parameters would converge to the same state, given the same initial conditions, for any random ordering of the time series. This is not the case here. Once vectorized and viewed as a regression problem, where $\chi(n-1)$ is the input and $x(n)$ is a desired signal, the input and desired signals can be randomly sorted, and the algorithm will still converge to the same state, given the same initial conditions. However, the input vectors would no longer represent the same time series. In addition, the gate, being a multilayer perceptron, possesses a weak memory in the sense that it must learn to predict the *a priori* probability of the experts by learning from their performance over the entire data set.

Clearly the architecture has memory, but does it have sufficient memory to achieve a perfect segmentation? The answer is no for two reasons. First, there may be overlap between the return maps of the dynamical systems in signal space or, in the language of nonlinear dynamics, the manifolds of the various dynamical regimes may intersect in phase space. Second, measurement noise can "smear" manifolds that ordinarily would never intersect. Where there is overlap in signal space, the gate cannot perform perfect segmentation. In brief, the predictive experts have short-term memory that by itself is insufficient for performing long-term segmentation. Therefore, we propose to use the unsupervised mixture of experts in conjunction with the supervised segmentation algorithms, exactly as for the memoryless mixture model case. That is, the mixture of experts are trained to completion on the entire data set. The "active" mode at the beginning of the data set is then identified using the multiple-hypothesis SPRT. Then the multiple-hypothesis CUSUM test is used to perform sequential change detection and identification on the entire time series with the understanding that, once a change is detected, the CUSUM test is reset from the change point. The process log-likelihood ratio between modes i and j up to time n is

$$L_{ij}(n) = \log\left[\frac{\prod_{m=1}^{n} p(i, x(m)|\chi(m-1))}{\prod_{m=1}^{n} p(j, x(m)|\chi(m-1))}\right] = \sum_{m=1}^{n} \log\left[\frac{h_i(m)}{h_j(m)}\right] \quad (4.113)$$

Once again, this can be written in a recursive manner, $L_{ij}(n) = L_{ij}(n-1) + L_{ij}(n)$, exactly as in (4.22), where $l_{ij}(n) = \log[(h_i(n))/(h_j(n))]$ is the log-likelihood increment of the posterior probabilities, exactly as for the memoryless mixture model (4.92). Just as in the memoryless case, the change detection algorithms are applied to the log-likelihood increment calculated with the posterior probability ratios. However, the argument for ignoring the priors is less compelling here than it was for the memoryless case because the priors are a function of the data itself. For example, if $i*$ is the true current active process, the log-likelihood ratio increment is

$$l_{i*j}(n) = \log\left[\frac{h_{i*}(n)}{h_j(n)}\right]$$

$$= \log\left[\frac{p(x(n)|\chi(n-1),i*)}{p(x(n)|\chi(n-1),j)}\right] + \log\left[\frac{g_{i*}(\chi(n-1))}{g_j(\chi(n-1))}\right] \quad (4.114)$$

Taking the expectation with respect to $p(x(n)|\chi(n-1),i*)$ of the first term on the right leads to the asymmetric Kullback–Leibler divergence between the pdf's of two random *processes*. The expectation with respect to the second term is more difficult to characterize because $g_i(\chi(n-1))$ is mapped by a multilayer perceptron.

The generalization properties of neural networks suggest the possibility of segmenting and identifying a time series without the use of the models themselves. Since the gate ideally learns to predict the posterior probabilities, it should be possible to obtain a segmentation using the log-likelihood increment approximated by the priors only:

$$l_{ij}(n) \approx \log\left[\frac{g_i(\chi(n-1))}{g_j(\chi(n-1))}\right] \quad (4.115)$$

If the models are available, however, the only apparent advantage to this is speed of computation.

4.5.4 Other Approaches

What if the recent past of the time series does not provide any information about the current regime? The most obvious approach is to assume the mixing coefficients are constants independent of the time series, $P(k|\chi(n-1)) = P(k) \equiv g_k$, just as for the memoryless mixture model of Section 4.4. The mixture of experts algorithm is easily modified by replacing the neural network gate with static mixture coefficients. For lack of a better term, we call this a static mixture of experts (SMOE). During training, the mixture coefficients are updated at the end of each iteration as the average of the posterior probabilities, exactly as in (4.88). Static mixture coefficients are consistent with many production models, of which three possible are:

- Choose a regime k^* according to the prior probabilities $P(k)$ and then calculate the next sample as $x(n) = f(\chi(n-1);\theta_{k^*}) + e_{k^*}(n)$.
- Choose a regime k^* according to the prior probabilities $P(k)$ and then generate N samples by recursively iterating $x(n) = f(\chi(n-1);\theta_{k^*}) + e_{k^*}(n)$.
- Choose a regime k^* with uniform probability, and then generate $NP(k^*)$ samples by recursively iterating $x(n) = f(\chi(n-1); \theta_{k^*}) + e_{k^*}(n)$.

It should be clear that only the latter two production models are consistent with a piecewise stationary time series. Once again, this suggests that fitting a time series to the model will not by itself provide any segmentation information, and further analysis is required to determine if the switching is consistent with piecewise stationarity.

For a time series generated according to one of these production models, the gate in the mixture of experts cannot learn to predict the regimes using the same data that the experts have to predict the next sample. However, there is a way to modify the gated mixture of experts so that it is appropriate whether or not the regime is predictable from the same tapped-delay line the experts see. If the gate is given a *longer* memory depth than that of the experts, the regime can always be predicted from the past of the time series (Weigend et al. 1995). For example, say the time series is generated from several regimes that each use the last M samples to generate the next value, and the regimes are randomly switched according to one of the piecewise stationary production models described above. This implies that although the last M samples are sufficient to predict the next value, they provide absolutely no information about the current regime and an input-driven gate would be useless. However, within the last $M + 1$ samples there is clearly *some* information about the current regime because a prediction could be made *and* evaluated entirely within those $M + 1$ samples. Within a tapped-delay line of $M + 2$ samples, there will be still more information about the current regime. In essence, as the state-space embedding dimension increases, the trajectories overlap less and less, and the regime becomes more and more predictable. This, then, is the basis for allowing the gate a longer tapped-delay line than the experts. It must be emphasized, however, that such mappings may be *very* complex because the gate is essentially performing a task similar to all the experts.

On a related note, recall that the development in this section implicitly assumed that the memory depth of each regime's process was the same. If this is not the case, some experts may have less information than they need to predict the next sample, while others may have more, which can lead to excessive bias or variance, respectively, in the experts' predictions. Alternatively, each expert can have a tapped-delay line with a different memory depth (Bengio et al. 1996), which must be hard coded into the architecture prior to training, in the hope that each expert will find the process with the corresponding approximate memory depth. Another method would be to use an alternative delay line structure with an adaptable memory depth, such as the gamma (Principe et al. 1993) or Laguerre (Masnadi-Shirazi and Ahmed 1991) filters.

4.5.5 Simulations

For data set A, the experts made their prediction using only a single past value, but the gate was allowed to see two past values. The system was trained for 100 iterations of the EM algorithm. For each iterations, both the experts and gate were trained for multiple epochs until no further improvement was seen. The learning curve in Fig. 4.16 is characteristic of the EM algorithm, showing a slow but steady improvement. The return maps of the experts after training are shown in Fig. 4.17. It is important to understand that the correspondence between an expert and a regime has no particular significance because it is random, depending on the initial conditions of the experts and gate, and can change with each training run. The overall approximation to the true maps is, at best, fair. In particular, expert 4, which is trying to approximate regime 1, has also learned part of the map of regime 4. This is also evident in the gate output and posterior probabilities, shown in Fig. 4.18. Some evidence of piecewise stationary switching is evident in the posterior probabilities. The gate has only partially learned the posterior probabilities, indicating the difficulty of trying to predict the expert when the regime is randomly switched rather than as a function of the time series. Nevertheless, taking the testing data posterior probabilities as input to the

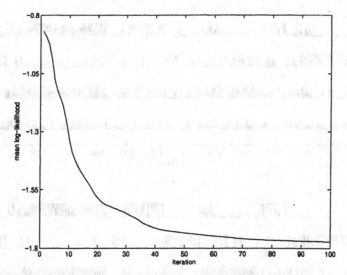

Figure 4.16 MOE, data set A, training: learning curve.

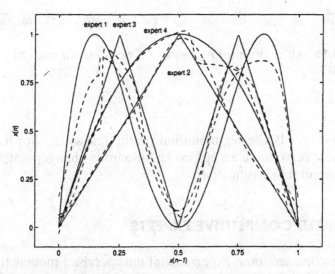

Figure 4.17 MOE, data set A, training: return maps of the experts (dashed line) and true return maps (solid line).

CUSUM test leads to the segmentation shown in Fig. 4.19. The segmentation is surprisingly good, but misses the change points at time 1078 and 1147. Also, a decision was incorrectly made in favor of regime 1 over regime 3 between times 507 and 607.

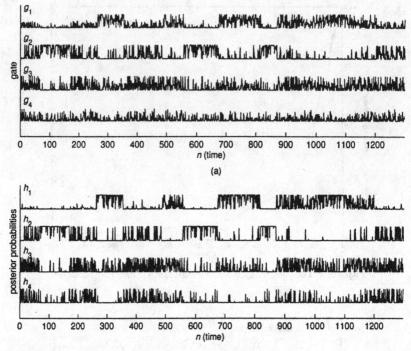

Figure 4.18 MOE, data set A, testing: (*a*) gate output and (*b*) posterior probabilities.

For data set B, the segmentation on the test set is shown in Fig. 4.20, and is reasonable except for an inaccurate changepoint, as well as an inaccurate classification.

4.6 GATED COMPETITIVE EXPERTS

We now consider some experimental unsupervised models that are no longer derived from the multivariate pdf of the time series but are a generalization of mixture models that fall within a single framework that we call *gated competitive experts*. Although some of the algorithms can be applied to memoryless processes, we will concentrate on processes with memory for the remainder of this chapter.

In this section we will develop a common framework for several seemingly disparate algorithms that have appeared in the literature recently, including the mixture of experts. In the following sections

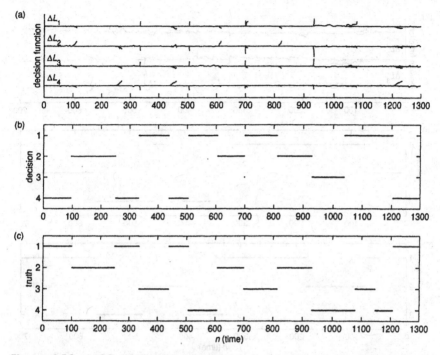

Figure 4.19   MOE, data set A, testing: (a) decision function applied to posterior probabilities, (b) decision, and (c) truth.

we will explore some of these algorithms. In particular, in the next section we will consider functional mappings other than prediction that can be used to provide segmentation information. Then, in the following section we will consider algorithms that can directly segment a time series, without resorting to an external segmentation algorithm, through the use of memory.

The astute reader will notice that there are no corresponding production models for these algorithms. As such, they should be considered experimental *analysis* tools whose utility can only be justified by their applicability to real data.

4.6.1 Overview

A gated competitive expert model is characterized by several adaptable "experts" that compete to explain the same data. That is, they all see the same input, and all attempt to produce the same output. Their performance is both monitored and mediated by a gate, whose

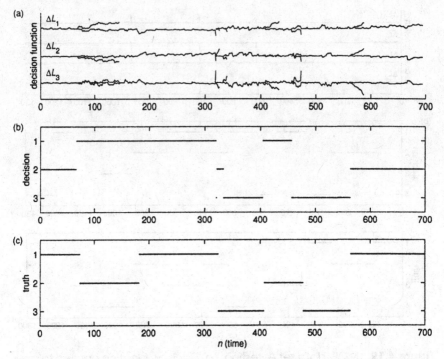

Figure 4.20 MOE, data set B, testing: (*a*) decision function applied to posterior probabilities, (*b*) decision, and (*c*) truth.

goal is to determine the relative validity of the experts for the current data, and then to appropriately moderate their learning. The history of the gate's decisions can be used to segment the time series.

In the activation phase, the total system output is a simple weighted sum of the experts' outputs

$$\mathbf{y}(n) = \sum_{k=1}^{K} g_k(n)\mathbf{y}_k(\boldsymbol{\chi}(n)) \qquad (4.116)$$

where $\mathbf{y}_k$ is the output of the kth expert with $\boldsymbol{\chi}$ as input, and g_k is the kth output of the gate. The input $\boldsymbol{\chi}(n)$ is a vector of delayed versions of a scalar time series, in essence the output of a tapped-delay line, repeated here for convenience:

$$\boldsymbol{\chi}(n) = [x(n)x(n-1)\ldots x(n-M+1)]^{\dagger} \qquad (4.117)$$

Note that we have allowed for the possibility that the kth expert's output, $\mathbf{y}_k$, may be a vector, even though we are still restricting the discussion to scalar time series. Why we do so shall become clearer shortly. In order for the mixture to be meaningful, the gate outputs are constrained to sum to one:

$$\sum_{k=1}^{K} g_k(n) = 1 \tag{4.118}$$

Such a linear mixture can represent either a competitive or cooperative system, depending on how the experts are penalized for errors, as determined by the cost function. In fact, it was in the context of introducing their mixture of experts model that Jacobs et al. (1991) first presented a cost function that encourages competition among gated expert networks, which we generalize here to

$$J(n) = \sum_{k=1}^{K} g_k(n) f(\mathbf{d}(n) - \mathbf{y}_k(\chi(n))) \tag{4.119}$$

where $\mathbf{d}$ is the desired signal and $f(\mathbf{d}(n) - \mathbf{y}_k(\chi(n)))$ is a function of the error between the desired signal and the kth expert. Since the desired signal is the same for all experts, they all try to regress the same data and are always in competition. This alone, however, is not enough to foster specialization. The gate uses information from the performance of the experts to produce the mixing coefficients. There are many variations of algorithms that fall within this framework. Let us discuss the important components one at a time, starting with the design of the desired signal.

4.6.2 Variations on the Theme

The formalism represented by (4.119) is a supervised algorithm in that it requires a desired signal. However, we are interested in a completely unsupervised algorithm. A supervised algorithm becomes unsupervised when the desired signal is a fixed transformation of the input: $\mathbf{d} \Rightarrow \mathbf{d}(\chi)$. Although many transformations are possible, the two most common transformations involve the delay operator and the identity matrix, resulting in *prediction* and *autoassociation*, respectively. We have already used prediction in conjunction with the mixture of experts in the previous section. In the next section, we will consider autoassociation.

Gates can be classified into two broad categories, which we designate as input or output based and are shown in Fig. 4.21. For output-based gating, the gate is a direct *calculated* function of the performance, and hence, the *outputs*, of the experts. The simplest

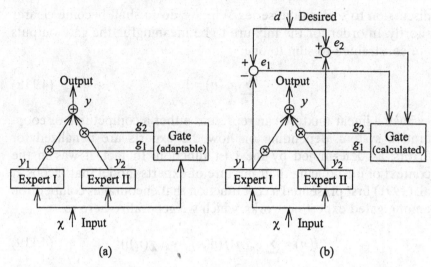

Figure 4.21 Two types of gates during activation: (*a*) input based and (*b*) output based.

scheme for the gate that falls within this category is hard competition, for which the gate chooses the expert with the smallest magnitude error in a winner-take-all fashion. Other output-based gates possess memory to keep track of past expert performance, the simplest example of which is the mixture model of Section 4.4 where the gate is the average of the posterior probabilities over the entire data set (4.88). The gate in the annealed competition of experts (ACE) (Pawelzik et al. 1996) implements memory in the form of a local boxcar average squared error of the experts. The self-annealing competitive prediction (Fancourt and Principe 1997) also uses the local squared error but uses a recursive estimator. Both of these algorithm will be discussed later.

With input-based gating, the gate is an *adaptable* function of the *input*, $g_k \Rightarrow g_k(\chi)$, that learns to forecast which expert will perform the best. The best example of input-based gating is the mixture of experts algorithm, which we have previously examined in some detail when the experts were predictors. In the next section we consider a mixture of experts type of algorithm when the desired signal is the input itself.

Finally, the total cost function can be constructed in several ways from the instantaneous cost, depending on the error function $f(\cdot)$. If

$f(\cdot)$ is a quadratic function of the error, the total cost over a data set of N samples is the sum of the instantaneous costs

$$J = \frac{1}{2}\sum_{n=1}^{N}\sum_{k=1}^{K}g_k(n)\|\mathbf{d}(n)-\mathbf{y}_k(\boldsymbol{\chi}(n))\|^2 \qquad (4.120)$$

If $f(\cdot)$ is a probability density function in the error, the appropriate total cost function is the negative log-likelihood

$$J = -\sum_{n=1}^{N}\log\left[\sum_{k=1}^{K}g_k(n)p(\mathbf{d}(n)-\mathbf{y}_k(\boldsymbol{\chi}(n)))\right] \qquad (4.121)$$

and returns us toward the mixture of experts when the gate is input based.

As pointed out by Jacobs et al. (1991), there is an important distinction between these two cost functions. Consider the instantaneous gradient of the cost function with respect to the ith weight in the kth expert, w_{ki}, for the case when $f(\cdot)$ is a quadratic function of the error

$$\frac{\partial}{\partial w_{ki}}J(n) = -g_k(n)[\mathbf{d}(n)-\mathbf{y}_k(n)]^\dagger\frac{\partial}{\partial w_{ki}}\mathbf{y}_k(n) \qquad (4.122)$$

and when $f(\cdot)$ is a Gaussian probability density function with error covariance $N(0,\sigma^2\mathbf{I})$:

$$\frac{\partial}{\partial w_{ki}}J(n) = -\frac{g_k(n)p_k(n)}{\sum_{j=1}^{K}g_j(n)p_j(n)}\frac{[\mathbf{d}(n)-\mathbf{y}_k(n)]^\dagger}{\sigma_k^2}\frac{\partial}{\partial w_{ki}}\mathbf{y}_k(n) \qquad (4.123)$$

For convenience we have assumed input-based gating so that $\partial g/\partial w_k$ = 0. Note that for a scalar desired signal, (4.123) is identical to the gradient in the mixture of experts algorithm (4.111). The primary difference between (4.122) and (4.123) is the probability term $p_k(n)$. Assuming for the moment that the gate assigns equal priors to all the experts, as would be appropriate early in the training, (4.122) tends to reward the *worst* performing expert, while (4.123) tends to reward the *best* performing expert. For this reason, specialization occurs faster with (4.123) than with (4.122). Unfortunately, it also means that in (4.123) the experts can begin to specialize without any significant contribution from the gate. On the other hand, in (4.122) the experts cannot begin to specialize without input from the gate, which thus has complete responsibility for specialization. This may be more conducive to annealing when it is required to recognize small changes in regime. The algorithms of Section 4.8 all employ annealing.

4.7   COMPETITIVE TEMPORAL PRINCIPAL COMPONENT ANALYSIS

As mentioned previously, any supervised algorithm can be converted to an unsupervised algorithm by making the desired signal a fixed transformation of the input. In this section we explore the case when the desired signal is the input itself:

$$J(n) = \sum_{k=1}^{K} g_k(n) f(\chi(n) - \hat{\chi}_k(n)) \qquad (4.124)$$

where $\hat{\chi}_k \equiv \mathbf{y}_k$ is the kth expert's estimate of the input.

Of course, any expert can immediately reconstruct the input by implementing an identity function. In order to place the experts in competition, they must be given a task that they cannot perform perfectly. One way to achieve this is for the experts to first reduce the dimensionality of the input and then attempt to reconstruct the input from the reduced dimensional space. The experts essentially compete on their ability to compress and then uncompress the input. Although there are many possible compression schemes, we will restrict ourselves to principal component analysis (PCA). PCA is a data-dependent lossy compression scheme that projects multidimensional data onto a subspace, creating an "information bottleneck" that must efficiently represent the statistics of the data in order to effect a good reconstruction.

We will fully develop the case of linear PCA experts first and later discuss the extension to nonlinear autoassociative experts. When linear PCA is applied to a vector of delayed versions of a time series, it is sometimes called the discrete Karhunen–Loeve transform; however, we prefer temporal PCA. It is not unreasonable to consider linear PCA first, since it has proven to be a useful tool in dynamical system theory for many years (Broomhead and King 1987).

In addition to the compression scheme, we also have to specify the gating structure. Dony and Haykin (1995) were the first to suggest competitive PCA in the context of image representation, and they used hard competition in the reconstruction error of the experts. In the nonlinear dynamics literature, local PCA based on neighborhoods has proven useful, which suggests input-based gating and the mixture of experts. We will utilize the nonlinear gated experts, where the gate is a nonlinear function of the input and $f(\cdot)$ is an appropriate pdf. In other words, we seek to combine linear PCA with the mixture of experts formalism. We first give a brief introduction to the terminology and notation and main results of PCA, without proof.

4.7.1 Temporal Principal Component Analysis

Consider expanding the zero mean vector $\chi(n)$, of dimension M, in terms of some fixed but complete basis in the M dimensional space

$$\chi(n) = \sum_{i=0}^{M-1} z_i(n)\mathbf{u}_i = \mathbf{U}z(n) \tag{4.125}$$

where the matrix $\mathbf{U} = [\mathbf{u}_0, \mathbf{u}_1 \ldots, \mathbf{u}_{M-1}]$ is deterministic and full rank. Let the basis be constrained to be orthonormal

$$\mathbf{u}_i^\dagger \mathbf{u}_j = \delta_{ij} \tag{4.126}$$

then, multiplying both sides of (4.125) from the left by $\mathbf{U}^\dagger$ immediately yields the coefficients of the expansion:

$$z_i(n) = \mathbf{u}_i^\dagger \chi(n) \Rightarrow z(n) = \mathbf{U}^\dagger \chi(n) \tag{4.127}$$

If the expansion is performed over a subset of the basis functions, the result is an estimate of χ:

$$\hat{\chi}(n) = \sum_{i=0}^{P-1} z_i(n)\mathbf{u}_i = \mathbf{U}_p z_p(n) \qquad P < M \tag{4.128}$$

where $\mathbf{U}_p$ is a matrix formed from a subset of the columns of $\mathbf{U}$. Likewise, the subset of the expansion coefficients is given by

$$\mathbf{z}_p(n) = \mathbf{U}_p^\dagger \chi(n) \tag{4.129}$$

Given a set of random vectors, $\chi(n), n = 1, \ldots, N$, what is the basis that minimizes the mean square reconstruction error over the data set, $E[\|\chi - \hat{\chi}\|^2]$, under the same orthonormality constraints as in

(4.126)? It is not difficult to show that the basis must satisfy the eigenvalue equation

$$\mathbf{R}\mathbf{u}_i = \lambda_i \mathbf{u}_i \qquad (4.130)$$

where λ_i and $\mathbf{u}_i$ are the ith eigenvalues and eigenvectors, respectively, of the autocorrelation matrix $\mathbf{R} = E[\chi\chi^\dagger]$. The P eigenvectors are chosen on the basis of the P largest corresponding eigenvalues. The projections z_i are called the principal components. The minimum mean square reconstruction error is given by the sum of the discarded eigenvalues

$$\min[E[\|\chi - \hat{\chi}\|^2]] = \sum_{i=P}^{M-1} \lambda_i \qquad (4.131)$$

The results up to this point are actually applicable to any random vector input. However, when the input χ is formed from delayed versions of a time series, the result is the temporal PCA architecture shown in Fig. 4.22, which will be a single expert within the context of a gated competitive system.

4.7.2 Temporal PCA in the Frequency Domain

When viewed as a set of FIR filters, referred to as eigenfilters, the eigenvectors of the autocorrelation matrix have very interesting properties. Szegö (see Thomson 1982) showed that the eigenvalues are asymptotically equal to the power spectrum sampled at equal

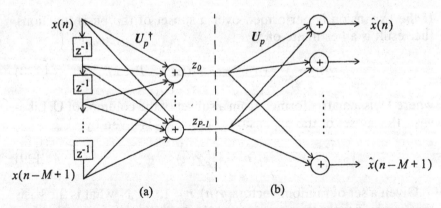

Figure 4.22 Temporal PCA architecture: (*a*) PCA network and (*b*) reconstruction network.

frequency intervals. To illustrate this in the infinite case, consider the Fourier transform of the eigenvalue equation (4.130)

$$\lambda_i U_i(f) = \int_{-1/2}^{1/2} U_i(f') S(f') \frac{\sin \pi M(f - f')}{\sin \pi(f - f')} \, df' \qquad (4.132)$$

where $S(f)$ is the power spectrum associated with the input signal autocorrelation function, $U_i(f)$ is the Fourier transform of the ith eigenfilter, and M is the eigenfilter dimension as defined previously. As $M \to \infty$, the Dirichlet function becomes a Dirac delta function and (4.132) becomes

$$\lambda_i U_i(f) = U_i(f) S(f) \qquad (4.133)$$

the solution of which is

$$U_i(f) = \delta(f - f_i) \qquad \lambda_i = S(f_i) \qquad (4.134)$$

That is, the eigenfilters form a fixed Fourier basis, and the eigenvalues become the power spectrum. Since the autocorrelation matrix is real and symmetric, the eigenvalues are real, and we can enforce the constraint that the eigenfilters be real. Then, each eigenvalue will be doubly degenerate, associated with two real sinusoidal eigenvectors separated in phase by 90 degrees. Assuming that the eigenvalues have been ordered from largest to smallest, so that λ_0 is the largest eigenvalue, then λ_0 is the maximum value of the power spectrum: $\lambda_0 = \max_f[S(f)]$.

For a finite-dimensional autocorrelation matrix the basis is no longer fixed but depends on the signal itself. However, as implied by Szegö's theorem, for a finite but sufficiently large eigenfilter dimension, M, the eigenfilters act as bandpass filters centered around peaks in the signal power spectrum, subject to orthogonality constraints. The passband becomes increasingly narrow as the number of taps is increased. Thus, eigenfilters can be approximately regarded as FIR filters matched to peaks in the signal spectrum. We can estimate the bandwidth by noting that since **R** is Toeplitz, the largest eigenfilter has all its zeros on the unit circle (Makhoul 1981) and thus forms the passband through the suppression of a zero near the peak frequency. Thus, the bandwidth is approximately $\Delta f \approx M^{-1}$.

The implications of this for using PCA to segment a time series is now clear. Recall that for Gaussian linear systems there is a one-to-one correspondence between stationarity and constancy of the spectrum. However, unlike the prediction task, which models the *entire* spectrum, linear PCA models the largest energy *peaks* in the spectrum. Thus it is possible to have a nonstationary signal for which the principal eigenfilters remain constant. However, there may be cases when such a feature is an advantage. For example, if there is additive noise in the signal, the principal eigenfilters will model the signal while the other eigenfilters will attempt to model the noise. Thus, by using only the first few eigenfilters, we can create a segmentation algorithm that is less sensitive to changes in the spectral properties of additive noise. For signals created by nonlinear systems, this convenient interpretation in terms of spectrum is lost. In this case, we just consider any PCA to be an experimental analysis tool.

4.7.3 The pdf of Reconstruction Error

In attempting to integrate PCA with the MOE, we start with the statistical mixture model view (4.119) of the MOE:

$$p(\mathbf{d}|\boldsymbol{\chi}) = \sum_{k=1}^{K} p(k|\boldsymbol{\chi})p(\mathbf{d}|\boldsymbol{\chi}, k) \equiv \sum_{k=1}^{K} g_k(\boldsymbol{\chi})p_k(\mathbf{d}|\boldsymbol{\chi}) \qquad (4.135)$$

where the kth gate output, $g_k(\boldsymbol{\chi})$, represents the *a priori* probability of the kth expert and $p_k(\mathbf{d}|\boldsymbol{\chi}) \equiv p(\mathbf{d}|\boldsymbol{\chi}, k)$ is the conditional pdf of the desired signal, given the input and the parameters of the kth expert. It is tempting to try and model $p_k(\mathbf{d}|\boldsymbol{\chi})$ as a Gaussian distribution in the error between the desired signal and the kth expert's output, which for autoassociation becomes

$$p_k(\mathbf{d}|\boldsymbol{\chi}) = \frac{1}{(2\pi)^{M/2}|\Sigma_k|^{1/2}} \exp\left[-\frac{1}{2}[\boldsymbol{\chi} - \hat{\boldsymbol{\chi}}_k]^\dagger \Sigma_k^{-1}[\boldsymbol{\chi} - \hat{\boldsymbol{\chi}}_k]\right] \qquad (4.136)$$

where Σ_k is the weighted covariance matrix of the reconstruction error of the kth expert. Equation (4.136) is perfectly valid when the experts are simple autoassociators (linear or nonlinear). However, as we will now show, for linear PCA experts Σ_k is not full rank and thus not invertible.

Consider a single PCA system with a matrix of eigenvectors $\mathbf{U} = [\mathbf{U}_p \ \mathbf{U}_s]$, where $\mathbf{U}_p$ is the first P principal eigenvectors associated with the corresponding largest eigenvalues, and $\mathbf{U}_s$ is the remaining S secondary eigenvectors associated with the corresponding smallest eigenvalues. Noting that $UU^\dagger = U_pU_p^\dagger + U_sU_p^\dagger = I$, the reconstruction error of an input vector can be written as $\chi(n) - \hat{\chi}(n) = [I - U_pU_p^\dagger]x(n) = U_sU_p^\dagger x(n)$, and the covariance of the reconstruction error is given by

$$\Sigma = E\left[(\chi - \hat{\chi})(\chi - \hat{\chi})^\dagger\right] = U_sU_s^\dagger RU_sU_s^\dagger = \sum_{i=P}^{M-1} \lambda_i u_i u_i^\dagger \quad (4.137)$$

Thus, the rank of the covariance matrix is given by the number of secondary eigenvectors, $S = M - P$. Since the covariance matrix is square and of dimension P, it is not full rank for any dimensionality reduction. Ultimately, all such direct attempts at formulating PCA in a statistical framework fail because *PCA is not a model for the data but merely a transformation of the data based on a decomposition of the covariance matrix.* This is an undesirable situation because we would like to use the same statistical framework for both linear PCA and nonlinear autoassociation. One way around this is to borrow some results from factor analysis [see Morrison (1976) or Jolliffe (1986)]. Factor analysis is a *model* that attempts to explain an observed high-dimensional random vector in terms of an unobserved lower-dimensional random vector. It can be shown that, under certain simplifying assumptions, factor analysis defaults to a PCA expansion. Factor analysis suggests that, for the linear PCA case, instead of modeling the reconstruction error, as in (4.135), we should directly model the pdf of the input as the weighted sum of the conditional pdf of the input, given the *K* PCA models:

$$p(\chi) = \sum_{k=1}^{K} g_k(\chi)p(\chi|k) \quad (4.138)$$

However, Fancourt and Principe (1998) show that the conditional probability of the input given the PCA parameters of the *k*th expert can be approximated as

$$p(\chi|k) \approx \frac{1}{(2\pi\sigma_k^2)^{S/2}} \exp\left[-\frac{\|\chi - \hat{\chi}_k\|^2}{2\sigma_k^2}\right] \quad (4.139)$$

where the variance is given by the mean value of the secondary eigenvalues

$$\sigma_k^2 = \frac{1}{S}\sum_{i=P}^{M-1} \lambda_{ki} = \frac{1}{S}E\left[\|\chi - \hat{\chi}_k\|^2\right] \quad (4.140)$$

and λ_{ki} is the ith eigenvalue of the kth expert. We will see shortly how to determine the eigenvectors and eigenvalues of the experts. Note that the approximate factor analysis model, consisting of (4.138), (4.139), and (4.140), is equivalent to the model based on reconstruction error in (4.135) and (4.136) if we set the covariance of the reconstruction error to have the diagonal form

$$\Sigma_k = \sigma_k^2 \mathbf{I} \tag{4.141}$$

and assume S degrees of freedom. Thus, using (4.141), we can use the same reconstruction error formulation, independently of whether the experts are linear or nonlinear. It is natural then to question the consequences of not using the exact pdf in (4.139). First, even the exact formulation makes a Gaussian assumption, which may be violated by the data set itself. Second, any approximation in the conditional pdf's of the experts can be accommodated by a small change in the mixture coefficients. Third, as we shall soon see, the approximation leads to a weighted mean square error derivation of the linear PCA parameters similar to the derivation for a single PCA system. Finally, the model works sufficiently well in practice.

4.7.4 Training the Model

We now have all the pieces required to train the model, given a time series, $x(n), n = 1, \ldots, N$. Again, we will use the EM algorithm, but this time we skip a detailed description, since the principles are very much the same as in Section 4.5. In the E step, for an initial set of model parameters, the posterior probabilities are evaluated using

$$h_k(n) \equiv \frac{g_k(\chi(n)) p(\chi(n)|k)}{\sum_{k=1}^{K} g_k(\chi(n)) p(\chi(n)|k)} \tag{4.142}$$

where $p(\chi|k)$ is given by the approximation in (4.139). In the M step, the cost function

$$J \approx \sum_{n=1}^{N} \sum_{k=1}^{K} \left\{ -h_k(n) \log[g_k(n)] \right.$$
$$\left. + \frac{1}{2} h_k(n) \left[\frac{\|\chi(n) - \hat{\chi}_k(n)\|^2}{\sigma_k^2} + S \log[\sigma_k^2] \right] \right\} \tag{4.143}$$

is globally minimized over the free parameters. We have written this as an approximation because the input vectors, $\chi(n)$, are *not* i.i.d. as a result of being delayed versions of the time series.

We now minimize (4.143) with respect to the free parameters. We skip intermediate steps, but see Fancourt and Principe (1998) for a more complete development. Recall that if the gate is a multilayer perceptron, the error back propagated to the input side of the softmax is given by (4.108), and (4.143) is usually only decreased and not minimized. However, since PCA is a linear transformation, we can globally minimize the second part of (4.143) with respect to the experts' weights. Furthermore, since the experts have been decoupled in (4.143), we can do this for each expert separately. Let the weights of the kth expert be given by some unknown $P \times M$ matrix $\mathbf{W}_k$ so that the reconstruction error is given by

$$\chi(n) - \hat{\chi}_k(n) = [\mathbf{I} - \mathbf{W}_k \mathbf{W}_k^\dagger]\chi(n) + \mathbf{b}_k \tag{4.144}$$

where we have included the unknown constant vector $\mathbf{b}_k$ to allow for the possibility that χ is not zero mean. Then the cost function of the kth expert is

$$J_k = \frac{1}{2}\sum_{n=1}^{N} h_k(n)\|\chi(n) - \hat{\chi}_k(n)\|^2 = \frac{1}{2}\sum_{n=1}^{N} h_k(n)$$
$$\{\chi^\dagger(n)[I - \mathbf{W}_k\mathbf{W}_k^\dagger]\chi(n) + 2\chi^\dagger(n)[I - \mathbf{W}_k\mathbf{W}_k^\dagger]\mathbf{b}_k + \mathbf{b}_k^\dagger\mathbf{b}_k\} \tag{4.145}$$

Finding the constant $\mathbf{b}_k$ first,

$$\mathbf{b}_k = -[\mathbf{I} - \mathbf{W}_k\mathbf{W}_k^\dagger]\bar{\chi}_k \tag{4.146}$$

where $\bar{\chi}_k$ is the mean of χ weighted by the posterior probabilities for expert k:

$$\bar{\chi}_k \equiv \frac{1}{\sum\limits_{n=1}^{N} h_k(n)} \sum_{n=1}^{N} h_k(n)\chi(n) \tag{4.147}$$

and $\mathbf{R}_k$ is a weighted autocorrelation matrix:

$$\mathbf{R}_k \equiv \frac{1}{\sum\limits_{n=1}^{N} h_k(n)} \sum_{n=1}^{N} h_k(n)[\chi(n) - \bar{\chi}_k][\chi(n) - \bar{\chi}_k]^\dagger \tag{4.148}$$

The jth column of the matrix $\mathbf{W}_k$, call it $\mathbf{w}_{kj}$, turns out to be an eigenvector of $\mathbf{R}_k$:

$$\mathbf{R}_k \mathbf{w}_{kj} = \lambda_{kj} \mathbf{w}_{kj} \qquad k = 1, \ldots, K \qquad j = 1, \ldots, P \qquad (4.149)$$

but chosen so that the eigenvectors correspond to the largest eigenvalues. That is to say, for the optimal reconstruction of the kth expert at each M step, we should choose the P largest eigenvectors of $\mathbf{R}_k$. This clearly shows that the competitive PCA architecture is still solving an eigenvalue problem. Finally, at each iteration, the variance of each experts' pdf is set to the average of the discarded eigenvalues, according to (4.140).

After training, the total system output can be considered to be either the principal components or the reconstruction of the input, weighted by gate:

$$\mathbf{z} = \sum_{k=1}^{K} g_k(\chi) \mathbf{z}_k(\chi) \qquad \tilde{\chi} = \sum_{k=1}^{K} g_k(\chi) \hat{\chi}_k(\chi) \qquad (4.150)$$

Once again, the gate only provides local segmentation information. Thus, the posterior probabilities can be used in conjunction with the supervised segmentation algorithms of Section 4.2, exactly as was done for the mixture models of Sections 4.4 and 4.5.

4.7.5 Other Approaches

Tipping and Bishop (1999) have also developed a mixture of linear PCA experts that they call mixtures of probabilistic principal component analyzers (MPPCA). There are two primary differences between their approach and ours. First, their mixture coefficients are constants, like the memoryless mixture model of Section 4.4, instead of being produced by an input-dependent gate. Second, they use the exact pdf of the input, instead of the approximation based on reconstruction error. This results in a mixture model of the form

$$p(\chi) = \sum_{k=1}^{K} g_k p(\chi|k) \qquad (4.151)$$

where $p(\chi|k)$ is a Gaussian distribution with mean $\bar{\chi}_k$ and covariance $\Sigma_k = \sigma_k^2 \mathbf{I} + \mathbf{W}_k \mathbf{W}_k^\dagger$, with $\{ \bar{\chi}_k, \sigma_k^2, \mathbf{W}_k \}$ being a set of free parameters for each expert that are fit to the data using the EM algorithm. Note that the exact pdf cannot be expressed solely as a function of reconstruction error.

Our goal in formulating the pdf of the PCA experts in terms of reconstruction error was to allow the same formulation to be used for linear PCA and nonlinear autoassociation. Perhaps the simplest implementation of nonlinear autoassociation (Kramer 1991) is to replace both the linear compression and reconstruction layers with a two-layer network consisting of a nonlinear layer followed by a linear layer. The desired signal is simply the input itself and the back-propagated error is the reconstruction error. Just as for the linear case, in order to create data compression there must be an information bottleneck in the middle of the neural network, which is accomplished by making the middle linear layer of smaller dimension than the input. However, unlike the linear PCA case, where the weight sharing between the compression and reconstruction stages causes the covariance of the reconstruction error to be less than full rank, the reconstruction error can be modeled as a full multivariate Gaussian distribution.

Finally, within the nonlinear dynamics community, local PCA has been used for calculating attractor dimension (Hediger et al. 1990), noise reduction (Sauer 1992), and prediction (Sauer 1994). These techniques utilize PCA performed within contiguous neighborhoods in state space where the neighborhoods are defined by various heuristics. By contrast, gated competitive PCA chooses the neighborhoods based on a competition in the reconstruction and, if the gate is a multilayer perceptron, the neighborhoods need not be contiguous.

We do not present any simulations for competitive linear PCA, however, because we did not get good results on either data set. Once we trained a mixture of linear PCA experts to completion, we then performed the CUSUM test on the posterior probabilities. As we saw in Section 4.5, the CUSUM test is very sensitive to even the slightest tendency of an expert to capture an extended temporal region. However, we saw no evidence of this. From this standpoint, we would have to conclude that linear PCA coefficents do not represent a unique property that can be associated with any of the regimes of either data set. Perhaps this can be best explained in the frequency domain. Temporal PCA is sensitive to peaks in the spectra, and its utility in segmenting signals with peaky spectra was demonstrated in Fancourt and Principe (1998). However, many chaotic signals are known to have spectral properties similar to broadband noise, which renders temporal PCA useless. Nevertheless, it remains an open question whether nonlinear PCA experts would provide better results for chaotic signals.

4.8 OUTPUT-BASED GATING ALGORITHMS

We have already discussed how the output of the gate does not provide long-term segmentation information, and for this reason the unsupervised mixture models must be used in conjunction with supervised change detection algorithms, resulting in a modeling–segmentation–modeling iteration. We now consider architectures that are capable of segmenting signals without requiring the use of an external segmentation algorithm. Clearly, such architectures will require additional memory over and above that encapsulated in the experts, in order to monitor the longer term performance of the experts. Through the use of memory, an expert that has performed well in the recent past can be given an advantage in the present. In general, memory can be added either to the gate or the experts themselves. Weigend et al. (1995) briefly discuss adding memory to the gate in the mixture of experts in the form of a time-delayed feedback of the gate outputs. Here, we restrict ourselves to output-based gates which, if you will recall, calculate the relative validity of the experts based solely on their recent performance.

4.8.1 Local Performance Measures

Recall that in Section 4.2.8, we found an expression for the probability of the ith predictive model given all the data up to the present time. However, rather than using all prior data, which may encapsulate multiple change points, a better approach is to restrict the calculation to the past τ time steps:

$$P(i|\text{recent past}) = \frac{\displaystyle\prod_{m=n-\tau+1}^{n} p_{e_i}(x(m) - f_i(\chi(m-1)))}{\displaystyle\sum_{k=1}^{K} \prod_{m=n-\tau+1}^{n} p_{e_k}(x(m) - f_k(\chi(m-1)))} \quad (4.152)$$

Note that there need not be any relationship between the implicit embedding dimension associated with the predictors and the number of time steps used to calculate the probabilities. If the residuals are all Gaussian, then this can be written

$$P(i|\text{recent past}) = \frac{\exp\left[\dfrac{-1}{2\sigma_i^2} \displaystyle\sum_{m=0}^{\tau-1} e_i^2(n-m)\right]}{\displaystyle\sum_{k=1}^{K} \exp\left[\dfrac{-1}{2\sigma_k^2} \displaystyle\sum_{m=0}^{\tau-1} e_k^2(n-m)\right]} \quad (4.153)$$

where e_i is the prediction error of the ith predictive model with variance σ_i^2. This suggests that model probability is a function of the local mean-squared-error

$$\varepsilon(n) = \frac{1}{\tau} \sum_{m=0}^{\tau-1} e^2(n-m) \tag{4.154}$$

This can also be estimated recursively

$$\varepsilon(n) = \gamma e^2(n) + (1-\gamma)\varepsilon(n-1) \qquad 0 < \gamma < 1 \tag{4.155}$$

where the parameter γ controls the memory depth of the filter. There are several ways to define the effective memory of a recursive system, but the simplest is to calculate the average of the temporal index weighted by the impulse response (Principe et al. 1993). According to this definition, it is easy to show that the effective memory depth of (4.155), in samples, is given by γ^{-1}. Furthermore, taking the expected value of both sides of (4.155) shows that it is an unbiased estimate of the mean square error in the steady state: $E[\varepsilon] = E[e^2]$.

For an on-line adaptive system, the change of some adaptable weight, w, is proportional to the instantaneous gradient

$$\frac{\partial}{\partial w}\varepsilon(n) = 2\gamma \frac{\partial}{\partial w}e(n) + (1-\gamma)\frac{\partial}{\partial w}\varepsilon(n-1) \tag{4.156}$$

Viewed in this way, (4.156) is equivalent to *learning with momentum*. In the neural network literature, momentum learning is presented as an *ad hoc* way of speeding up learning. Here, we see it is the natural consequence of a recursive mean-square-error cost function. We now examine ways to incorporate these local performance measures into the gate.

4.8.2 Annealed Competition of Experts

Perhaps the most successful example of output-based gating is the annealed competition of experts (ACE) algorithm (Pawelzik et al. 1996). A set of K expert neural predictors operate in parallel, and a gating function determines the local-in-time probabilities of the experts. The gating function is formed from (4.153) under the assumption that all the experts have equal prediction error variance

$$g_k(n) = \frac{e^{-\beta \varepsilon_k(n)}}{\sum\limits_{j=1}^{K} e^{-\beta \varepsilon_j(n)}} \qquad (4.157)$$

where β is a constant that has absorbed the prediction error variance, and ε_k is the local sum-of-squared error of the kth expert, calculated using a *noncausal* boxcar filter

$$\varepsilon_k(n) = \sum_{m=-\tau}^{\tau} e_k^2(n-m) \qquad (4.158)$$

and e_k is the prediction error of the kth expert.

The total cost function is based on the gate-weighted sum-of-squared error, exactly as in (4.120). The gradient of the total cost function with respect to the ith weight, w_{ki}, in the kth expert is calculated by treating the gate as though it were a constant:

$$\Delta w_{ki} = -\eta \frac{\partial J}{\partial w_{ki}} = -\eta \sum_{n=1}^{N} g_k(n) e_k(n) \frac{\partial}{\partial w_{ki}} y_k(n) \qquad (4.159)$$

where η is the learning rate. Equation (4.159) suggests a batch approach, but the original authors used a sample-by-sample update, in effect dropping the summation. The gate output can then be viewed as a weighting term to the learning rate of the individual experts.

The key to the ACE algorithm is the interpretation of the β parameter. Rather than determining β from the predictors' performance, it is instead viewed as an externally adjustable parameter that determines the degree of competition. It is then used to anneal the competition during training from a soft toward a hard competition. By starting the learning process with a small value of β, all the experts identically see the entire data set and converge to an average of all the dynamical regimes. Then as β is increased and the competition becomes harder, each predictor begins to specialize on a subset of regimes, which then puts it at a disadvantage for other regimes, allowing other predictors to win those regions. Pawelzik et al. (1996) report *phase transitions*, specific values of β for which a dramatic change in the cost occurs as the experts begin to specialize. There may be several such phase transitions in the course of training. As $\beta \to \infty$, the competition becomes hard and only the winning predictor earns the right to update its prediction, at which point the experts are fine-tuning their modeling of their respective regimes. After each change in β, it is important to train the experts to completion before any further annealing.

The other key issue is the memory term, in the form of a lowpass "boxcar" filter, for the local sum-of-squares error (4.158). In cases where there is a large overlap between the return maps of the various dynamical regimes, or a small signal-to-noise ratio, a single time instance is insufficient to estimate the probabilities of the experts. This can result in rapid switching between predictors. However, under the assumption of a relatively slow switching rate, a single predictor would be expected to win for a longer time period. Increasing the memory depth gives a greater advantage to the predictor that has won in the recent past. The memory parameter τ is chosen beforehand and fixed at that value during training. Note that by using the local sum-of-squares error, there is a coupling between the competition and the memory depth. However, this is not a factor if the memory depth is not changed during training. The acausal nature of the filter is not a hindrance because of the off-line nature of the algorithm, but it ensures that regime switches indicated by the gating function are aligned with the data.

4.8.3 Simulations

Our implementation of the ACE algorithm differs from the authors in several respects. First, we use batch learning, whereas the original authors update the weights at each time step. Second, we use the local mean squared error, rather than the local sum-of-squares error, in order to completely decouple the memory depth from the competition. This does not affect the training, but allows for experimenting with different memory depths without having to adjust the annealing parameter.

For data set A, the memory depth was chosen to be 8 based on experimentation. The competition parameter β was linearly annealed from 1 to 300 during the 300 epochs of training. The log of the gate-weighted mean square error during training is shown in Fig. 4.23, for which three distinct phase transitions are clearly visible. Hard competition was then invoked and the MLPs were trained again to completion. The return maps of the four experts after training are shown in Fig. 4.24. Once again, it is important to recognize that the correspondence between an expert and a regime is completely random, depending on the initial conditions of the neural

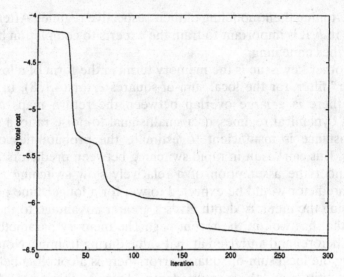

Figure 4.23   ACE, data set A, training: learning curve.

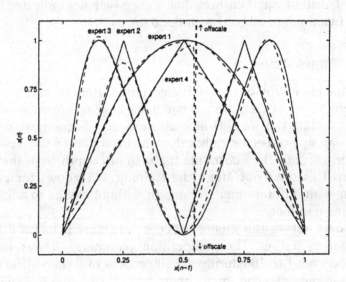

Figure 4.24   ACE, data set A, training: experts' return maps (dashed) and true return (solid).

networks. Although the return maps are good approximations to the true maps, there is an anomaly in the return map of expert 4 due to excessively large weight values caused by overtraining. This occurred repeatedly over several training runs. We could have used some sort of regularization technique during training to eliminate this effect but left it as is in order to make the point that overtraining can be a problem due to the large number of times a network must be trained during the annealing schedule. This anomaly caused minor segmentation errors on the training set and also shows up on the test set shown in Fig. 4.25. Pawelzik et al. (1996) used radial basis function experts, which are not vulnerable to this kind of overtraining. Nevertheless, the overall segmentation is still very good.

Data set B required more annealing. We trained for 1000 iterations and there was a phase transition as late as the 820th iteration. We simply present the results on the test set in Fig. 4.26. The results are good except for some anomalies that can again be attributed to overtraining.

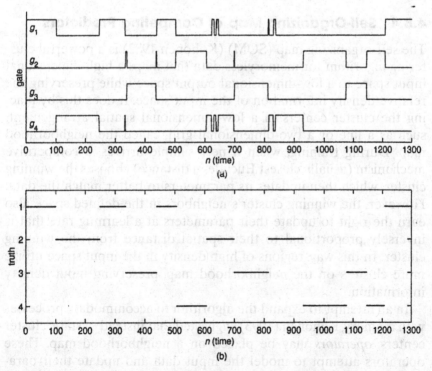

Figure 4.25 ACE, data set A, testing: (*a*) gate output and (*b*) truth.

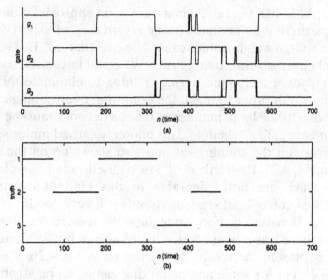

Figure 4.26 ACE, data set B, testing: (*a*) gate output and (*b*) truth.

4.8.4 Self-Organizing Map of Competing Predictors

The self-organizing map (SOM) (Kohonen 1982) is a powerful clustering algorithm for memoryless data that maps a high-dimensional input space to a low-dimensional output space while preserving the relative density information of the input space. It does this by placing the cluster centers in a low-dimensional spatial arrangement, such as a line or a two-dimensional grid, called the neighborhood map. During training, when a new sample arrives, a competitive mechanism (usually closest Euclidean distance) chooses the winning cluster, which then updates its parameters to better match the data. However, the winning cluster's neighbors in the defined space also earn the right to update their parameters at a learning rate that is inversely proportional to their spatial distance from the winning cluster. In this way, regions of high density in the input space utilize more clusters on the neighborhood map, preserving input density information.

In an attempt to expand the algorithm to accommodate processes with memory, Kohonen (1993) suggested that instead of static cluster centers, *operators* may be placed on a neighborhood map. These operators attempt to model the input data and update their parameters on the basis of their distance on a neighborhood map from

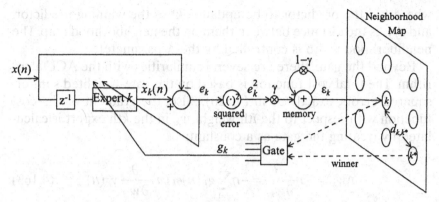

Figure 4.27 Neighborhood map of competing predictors.

the best model. Kohonen (1993), and independently Fancourt and Principe (1996), proposed using linear predictor operators and their prediction error as the competitive basis for choosing the winning model. This architecture is shown in Fig. 4.27, where the predictors have been placed on a two-dimensional grid. Here we extend this architecture to the use of nonlinear predictors.

In the self-organizing map of competing predictors (SOMCP) (Fancourt and Principe 1996), the winning predictor is the one with the smallest local mean squared error:

$$\text{Winner}(n) = \text{argmin}_k[\varepsilon_k(n)] \tag{4.160}$$

computed using the *recursive* estimate, as in (4.155):

$$\varepsilon_k(n) = \gamma e_k^2(n) + (1-\gamma)\varepsilon_k(n-1) \qquad 0 < \gamma \le 1 \tag{4.161}$$

where e_k is the prediction error of the kth expert. The memory term γ is identical for all experts.

This architecture fits in quite nicely with the framework outlined in Section 4.6. Here, the gating function is determined by the distance from the winning predictor to the other predictors:

$$g_k(n) = \frac{\exp\left[\dfrac{d_{k,k*}^2(n)}{2\Lambda^2(n)}\right]}{\displaystyle\sum_{j=1}^{K}\exp\left[\dfrac{d_{j,k*}^2(n)}{2\Lambda^2(n)}\right]} \tag{4.162}$$

where k is the predictor to be updated, k^* is the winning predictor, and d_{k,k^*} is the distance between them on the neighborhood map. The neighborhood width is controlled by the Λ parameter.

Beyond the gate, there are several similarities with the ACE algorithm. The total cost function is based on the gate-weighted sum-of-squared errors, exactly as in (4.120). Also, the gradient of the cost function with respect to the ith weight w_{ki} in the kth expert is calculated by treating the gate as a constant:

$$\Delta w_{ki} = -\eta \frac{\partial J}{\partial w_{ki}} = -\eta \sum_{n=1}^{N} g_k(n)e_k(n)\frac{\partial}{\partial w_{ki}}y_k(n) \qquad (4.163)$$

Although the weights can be updated on-line, we prefer the batch approach indicated in (4.163), in which case the gate output can be viewed simply as a weighting factor for the data.

In exact analogy with training a self-organizing map, the neighborhood width is annealed during training according to an exponentially decreasing schedule

$$\Lambda(i) = \Lambda_0 \exp\left[\frac{-i}{\alpha}\right] \qquad (4.164)$$

where α is an annealing rate and i is the iteration number. Early in the training when the neighborhood width is large, all the experts essentially see the same data and converge to an average of the dynamics in the data. As the neighborhood width is decreased, the experts begin to specialize, but with similar dynamical systems appearing as neighbors on the map. The memory term need not be annealed, but our experience is that doing so generally yields better results. We anneal it according to a linear schedule

$$\gamma(n) = 1 - \beta i \qquad (4.165)$$

where β is an annealing rate.

Just as in a static SOM, there are several possible choices for the map structure. The map is almost always chosen to be one or two dimensional so that results can be easily visualized. For a one-dimensional map, the experts are placed equidistant on a line, the ends of which can also be made neighbors to form a circle. For a two-dimensional map, the experts are placed on a square grid, the corners of which can be made neighbors to form a torus.

The SOMCP seems particularly adept at segmenting cyclo-stationary processes, where a system is only periodically excited. When inverse filtered through a predictor, the excitation cannot be predicted, and the result is periodically large prediction errors. These large errors tend to distort the competition process of the other algorithms. However, since the competitive process of the SOMCP is only based on the distance to the winning predictor, the relative reward system remains unchanged, independent of the actual performance difference between the winning predictor and the other predictors.

4.8.5 Simulations

For data set A, four neural network experts were placed on a line map and trained for 200 iterations. Both the neighborhood width and memory depth were annealed as shown in Fig. 4.28b. At each iteration, the neural networks were trained for several epochs using the most recent gate information. The learning curve is shown in Fig. 4.28a, which does not exhibit phase transitions like in the ACE algorithm but learns equally well, as evidence by the return maps shown in Fig. 4.29. Unlike the previous algorithms, however, there *is* a correspondence between the experts and the regimes. That is, every run

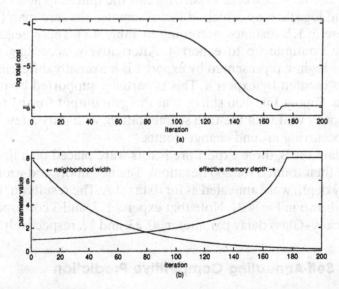

Figure 4.28 SOMCP, data set A, training: (a) learning curve and (b) annealing schedule.

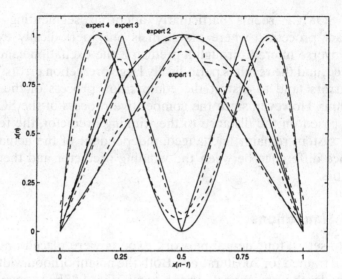

Figure 4.29 SOMCP, data set A, training: experts' return maps (dashed) and true return (solid).

produces the same alignment of experts with regimes (although the order may be reversed). Furthermore, experts that are neighbors on the neighborhood map tend to grab regimes that are similar. This can be seen in Fig. 4.29 where expert 1 grabs the tent map and expert 2 takes the logistic map, which in fact resembles the tent map (and is also close in KL distance, according to Table 4.1). These neighborly relations continue up to expert 4. Alternatively, according to the map, the regime represented by expert 1 is maximally different from that represented by expert 4. This is partially supported by the KL distance. Finally, Fig. 4.30 shows that the gate output for the test set corresponds well with the true segmentation, with only a few minor errors occurring around change points.

For data set B, three expert predictors were placed on a line map and the then trained for 200 iterations. The neighborhood width and memory depth were annealed as for data set A. The results on the test set are shown in Fig. 4.31. Note that experts 1, 2, and 3 correspond to the Mackey–Glass delay parameter 30, 23, and 17, respectively.

4.8.6 Self-Annealing Competitive Prediction

Recall that the ACE algorithm uses an output-based gate with memory, in the form of a boxcar moving average (MA) of the local

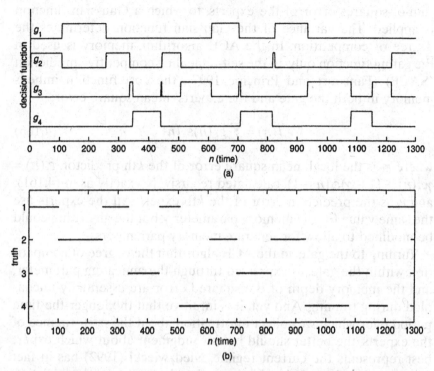

Figure 4.30 SOMCP, data set A, testing: (*a*) gate output and (*b*) truth.

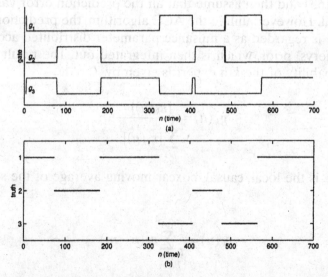

Figure 4.31 SOMCP, data set B, testing: (*a*) gate output and (*b*) truth.

sum-of-squares error of the experts, to which a Gaussian function is applied. The variance of the Gaussian function determines the degree of competition. In the ACE algorithm, memory is used in the gating function only. In the self-annealing competitive prediction (SACP) (Fancourt and Principe 1997), the cost function imbeds memory in both the gate and the experts' mean square error:

$$J(n) = \sum_{k=1}^{K} g_k(n)\varepsilon_k(n) \qquad (4.166)$$

where ε_k is the local mean square error of the kth predictor, $\varepsilon_k(n) = \gamma e_k^2(n) + (1 - \gamma)\varepsilon_k(n - 1)$, calculated recursively exactly as in (4.161), and e_k is the prediction error of the kth expert. All the experts use the same value for the memory parameter γ, but the algorithm could be modified to allow for separate memory parameters.

Turning to the gate, in the ACE algorithm the degree of competition within the gate, as expressed through the annealing parameter, and the memory depth of the squared error are essentially uncoupled during training. And yet, it is intuitive that the longer the time period over which we collect information about the performance of the experts, the better should be our judgment about which expert best represents the current regime. Niedzwiecki (1992) has in fact formulated such a relationship, assuming that the expert's prediction error variances are unknown but distributed according to a so-called noninformative prior distribution. Like the ACE algorithm, we begin with (4.153) and then assume that all the prediction error variances are equal. However, unlike the ACE algorithm, the prediction error variance is regarded as a nuisance parameter distributed according to a Jefferys' prior, which is then integrated out. The result is that the probability of the kth expert is given by

$$g_k(n) = \frac{[\varepsilon_k(n)]^{\frac{-\tau}{2}}}{\sum_{j=1}^{K}[\varepsilon_j(n)]^{\frac{-\tau}{2}}} \qquad (4.167)$$

where ε_k is the local, causal, boxcar moving average of the squared error

$$\varepsilon_k(n) = \frac{1}{\tau}\sum_{m=0}^{\tau-1} e_k^2(n-m) \qquad (4.168)$$

The gate in (4.167) embodies the coupling between memory depth and degree of competition. When $\tau = 1$, the unnormalized gate is just the inverse of the absolute value of the instantaneous error $g_k(n) \propto |e_k(n)|^{-1}$. As $\tau \to \infty$, the gate implements hard competition such that $g_k(n) = 1$ when k represents the expert with the smallest mean square error, and $g_k(n) = 0$ for all others.

Equations (4.167) and (4.168) were developed for evaluating the local performance of *known* experts. Here, we incorporate them into the cost function (4.166) for adaptable experts. We use (4.167) as the gating function, except that we replace the moving-average calculation of the mean square average in (4.168) by our recursive calculation in (4.161). Likewise, we replace the memory depth τ in (4.167) by the recursive memory parameter γ^{-1}, yielding the gating function

$$g_k(n) = \frac{[\varepsilon_k(n)]^{\frac{-1}{2\gamma}}}{\sum_{j=1}^{K}[\varepsilon_j(n)]^{\frac{-1}{2\gamma}}} \qquad (4.169)$$

We then do gradient descent on *both* the calculated gate and the mean square errors of the instantaneous gradient (4.166). Taking the partial derivative with respect to the ith weight w_{ki} in the kth expert, resulting in

$$\Delta w_{ki}(n) = -\eta \frac{\partial}{\partial w_{ki}} J(n) = -\eta g_k(n)\left[\frac{J(n)}{\varepsilon_k(n)} + 2\gamma - 1\right] z_{ki}(n) \quad (4.170)$$

$$z_{ki}(n) = e_k(n)\frac{\partial}{\partial w_{ki}} y_k(n) + (1-\gamma)z_k(n-1) \qquad (4.171)$$

and with respect to the γ parameter

$$\Delta\gamma = \frac{-\eta'}{2\gamma}\sum_{k=1}^{K} g_k(n)\left\{\left(\frac{J(n)}{\varepsilon_k(n)} + 2\gamma - 1\right)v_k(n)\right.$$
$$\left. - \frac{[J(n) - \varepsilon_k(n)]\log[\varepsilon_k(n)]}{\gamma}\right\} \qquad (4.172)$$

$$v_k(t) = \frac{\varepsilon_k(n) - \varepsilon_k(n-1)}{\gamma} + (1-\gamma)v_k(n-1) \qquad (4.173)$$

Equation (4.171) for z_k is the standard weight update equation *with momentum* for the kth expert operating independently. Thus, momentum learning falls out naturally as a result of using the mean square error in the cost function (4.166) instead of the instantaneous error. The competitive nature of the algorithm is evident in (4.170), where the total weight update is given by the product of z_k with the probability of the kth expert, g_k, and another term that depends on the inverse of the mean square error of the kth expert. Thus, the expert with the smallest mean square error will have the largest weight update. Furthermore, (4.170) also shows how the annealing is coupled with the memory depth. For $\gamma > 0.5$, the term in brackets is always positive, and thus all the experts get to improve their predictions. However, for $\gamma < 0.5$, the sign of the term in brackets can be either positive or negative, depending on whether the mean square error of the kth predictor is less or greater than, respectively, the total cost. Thus, experts that perform poorly can actually be pushed away from the data, although at a small learning rate due to the gating function.

4.9 OTHER APPROACHES

There are several other algorithms that either did not fit into the paradigm of gated competitive experts, were not appropriate for the subject of the chapter because they involve recurrent networks, or were beyond the scope of this chapter. For the sake of completeness, we briefly mention them here.

4.9.1 Hierarchical Mixture of Experts

We have not discussed the possibility that a signal may be composed of a mixture of mixtures. The mixture models presented in Sections 4.4 and 4.5 can readily be extended to a mixture of mixtures. For the memoryless case, one of several possible corresponding production models is

- Choose a *subset* of regimes labeled by the index j^* according to the prior probabilities $P(j)$, then choose a regime k^* according to the prior conditional probabilities $P(k|j^*)$, and then generate N-independent samples from the distribution $p_{k^*}(x)$.

In the case of the mixture of experts, the extension is called the hierarchical mixture of experts (HME) (Jordan and Jacobs 1994) and employs a hierarchy of simple logistic gating functions, each of which mixes the outputs from the previous level, the lowest level being the outputs of the experts themselves. The primary advantages of such an approach are fast training of the gate, the parameters of which can be found approximately in a single pass through the data. However, a single multilayer perceptron is a universal mapper, and therefore the gate in the nonlinear gated experts from Section 4.5 is sufficient for any problem that the HME can solve.

4.9.2 Hidden Control Neural Network

Another architecture useful for segmentation, but one that does not fit into the mixture model paradigm, is the hidden control neural network (HCNN) (Levin 1993). The HCNN uses a single multilayer perceptron to perform prediction but reserves one (or more) input as a hidden latent "control" variable, which can be either continuous or discrete. For the segmentation task, the control variable is usually a discrete variable. Training is accomplished by alternating between optimizing the neural network's weight set for a given control sequence and optimizing the control sequence for a given set of neural network weights. The HCNN is capable of modeling very fast regime switching, up to the maximal rate of one per time step. There is no guarantee, however, that the control variable always represents credible segmentation information, but the Jacobian (the derivative of the output with respect to the control variable) can be used to estimate its significance.

4.9.3 Drifting Dynamics

Not all changes between regimes may be abrupt, in which case the time series may be quasi-stationary. However, the dividing line between quasi-stationary and nonstationary is not exact, as evidenced by the debate over the nature of human speech. Nevertheless, Kohlmorgen et al. (1997) have devised a technique for detecting *deterministic* linear drifts between regimes (as opposed to a drift of model parameters). The search for drift is done *after* regular training with the ACE algorithm.

We now turn to algorithms that attempt to *statistically* model the transitions between regimes. Such higher level modeling requires a marked increase in the amount of training data in order to accurately characterize the statistics of the transitions between states. The training algorithms are also inevitably more complex and time consuming.

4.9.4 Hidden Markov Switching

We previously mentioned that the hidden Markov filter (Poritz 1982) and mixture autoregressive hidden Markov model (Juang and Rabiner 1985) assume switching between a finite set of linear autoregressive systems, where the switching between regimes is governed by a Markov chain. This approach has recently been extended to nonlinear models in an architecture called Markov gated experts (MGE) (Shi and Weigend 1997). The corresponding production model is

- Choose a new state k^* at time n based on the previous state j^* at time $n - 1$ according to the state transition probabilities $P(k|j^*)$ and then generate the next sample as $x(n) = f(\chi(n - 1); \theta_{k^*}) + e_{k^*}(n)$.

The resulting time series may be piecewise stationary (but not necessarily) if the probability of remaining in the current state, $P(k|k)$, is much greater than the probability of exiting the current state. Given a time series, the training problem is to estimate the state transition probabilities, $P(k|j)$, and optimize the parameters of the nonlinear predictors, θ_k, which are implemented using multilayer perceptrons. The EM algorithm can be used to estimate the hidden state sequence, and the resulting iterative training algorithm is similar to training a hidden Markov model, involving the so-called forward–backward procedure. Instead of discovering the state transition probabilities during training, it is also possible to simply fix them *a priori* to reflect typical transition rates between regimes (Kohlmorgen et al. 1999), greatly reducing the amount of training data required. In either case, the problem with using static transition probabilities is that they represent prior information about the expected switching rate between regimes which, if locally violated, can lead to segmentation and modeling errors. One way to counter this is to model the state durations as probability distributions, as is sometimes done in feature-vector-based HMM modeling. To the best of our knowledge, this has not yet been attempted in the context of Markov gated experts. Another possibility is to allow the transition probabilities to depend on the input, which leads to three closely related models, which we discuss next.

The input–output HMM (IOHMM) (Bengio and Frasconi 1996), the mixture of controllers (MOC) (Cacciatore and Nowlan 1994), and the hidden Markov mixtures of experts (HMME) (Kohlmorgen et al. 1999) are all essentially a combination of the mixture of experts and a hidden Markov model. In particular, the state transition probabilities depend on both the previous state (like an HMM or gated Markov experts) and the current input (like the mixture of experts). The corresponding production model is thus

- Choose a new state k^* at time n based on $\chi(n-1)$ and the previous state j^* at time $n-1$ according to the state transition probabilities $P(k|j^*,\chi(n-1))$ and then generate the next sample as $x(n) = f(\chi(n-1);\theta_{k*}) + e_{k*}(n)$.

In the MOC and IOHMM, the state transition probabilities are calculated using input-driven multilayer perceptron gates with recurrent connections. For the MOC the gate outputs feed back to the gate input, while for the IOHMM, the recurrent connections are at the output of the gate. While the HMME also has an input-driven gate, the recurrency is in the calculation of the posterior probabilities. Given a time series, the training problem is to optimize the parameters of the gate and the nonlinear predictors, θ_k, which are both implemented using neural networks. Once again, the EM algorithm is used to estimate the hidden state sequence, and the resulting training algorithm is similar to training a hidden Markov model.

To obtain segmentation information from a trained Markov switching model, it is entirely feasible to apply it to a time series (either the one it was trained on or a new one) and then use the CUSUM algorithm to perform the segmentation, exactly as we did for the mixture of experts. The only difference is that we must take into account the transition probabilities when calculating the log likelihood ratio increments. An alternative approach for Markov chains, however, is to find the most likely state sequence, which can be efficiently calculated using the Viterbi algorithm. The Viterbi algorithm chooses the most likely state at each time step. It can be regarded as a multiple-hypothesis SPRT where a decision is forced at each time step.

4.10 CONCLUSIONS

We began this chapter with the classical definitions of stationary random processes from early in the twentieth century and then discussed an important subset of nonstationary processes; namely, piecewise and quasi-stationary processes. A special subset of such processes is a time series produced by random switching among a finite set of dynamical systems, which we called regimes. Using the prediction error decomposition, we were able to express the likelihood of a random process relative to a multivariate pdf model in terms of the innovations of a predictive model. We then showed how to segment and model an unknown piecewise stationary time series using a variety of different unsupervised algorithms, all based on predictive models that we called experts. This opened the door to working with nonlinear processes and even ergodic chaotic signals, where the concept of stationarity breaks down, because chaotic signals are indistinguishable from other nonlinear signals for single-step prediction. We were then no longer strictly testing for changes in statistical stationarity but rather changes in the dynamics of the signal. However, from the measurement viewpoint of working with unknown data, this created a working definition of stationarity that depended on the power of the models we used.

We now review the important characteristics of the models presented, and suggest some future research directions.

4.10.1 Summary

We can separate the methods presented into three basic approaches. First, for the generalized likelihood ratio (GLR), the emphasis is on sequential detection of the change points. While it develops local-in-time models necessary for change detection, it does not attempt to build global models that represent the entire time series. In essence, it looks for changes in the local-in-time dynamics by training models on neighboring blocks of data and examining the likelihood ratio of the model's prediction error. Any increase in the likelihood ratio, which has a convenient interpretation in terms of Kullback–Leibler divergence, is a sign of a nonstationary change. The GLR is computationally intensive, and therefore it is necessary to make simplifications that allow it to be used with more complex models such as neural networks. It is not clear, however, which simplifications preserve the best performance.

Second, in the mixture model approach, the emphasis is on first developing models for the regimes over the entire time series, and then using those trained models to segment and identify the regimes within the time series using supervised techniques such as the multiple-hypothesis CUSUM test. For processes with memory, this led naturally to the mixture of experts (MOE). Although the MOE actually assumes a switching mechanism that depends on the time series itself, it can still be used even if this is not the case by allowing the gate more input memory than the experts. Alternatively, one can simply assume that the mixing coefficients are static constants. Unfortunately, when training either variation with the EM algorithm, the convergence results are less than satisfactory. While the EM algorithm guarantees a steady improvement in fitting the model to the data, it does so at the cost of going straight to the nearest local minima. It is still not clear if there are other training methods that will provide better results, such as second-order methods or annealing. Nevertheless, in simulations the experts were sufficiently specific to the regimes to allow a reasonable segmentation when teamed with the CUSUM test.

Third, methods such as the annealed competition of experts (ACE) and the neighborhood map of competing predictors (SOMCP) attempt to *simultaneously* segment and model the time series, by combining some of the elements of the sequential and mixture methods. Like the mixture models, they place the experts in competition over the entire time series. Like the sequential methods, they use the local-in-time performance of the experts for credit assignment. Unlike the MOE, however, the ACE and SOMCP do not model prediction error variance and thus may have difficulty with large variations in power. However, there is no obvious theoretical obstacle to including variance as a model parameter. Unlike the GLR, in simulations the ACE and the SOMCP performed very well without any modification for both the short-term memory of data set A and the longer term memory of data set B. However, there are still some important unanswered questions for these algorithms. For the CUSUM test, the relevant distance measure between regimes is the asymmetric KL divergence. Because it is asymmetric, it can be easier to detect a transition from A to B than from B to A. However, the ACE and the SOMCP do not differentiate between the order the regimes are presented. But might there be an appropriate *symmetric* distance measure that governs how well these algorithms can do?

We found that many of these approaches fit into a single framework that we called gated competitive experts. This framework allows for generalization in many new directions, one of which is the competitive principal component analysis (PCA), where the competition is based on the ability of the experts to compress and uncompress the data, rather than predict the next value of the time series. Competitive linear PCA was of no utility in segmenting switching chaotic signals, even when teamed with the CUSUM test. The PCA experts that arose out of competitive PCA bore no relationship to the chaotic regimes. We found this somewhat surprising at first, given PCA's widespread use in the nonlinear dynamics community. However, upon further reflection, while chaos tends to produce a spectra similar to broadband noise, linear PCA can only model "peaky" spectra. Does this imply that finding the principal components of a chaotic signal is a meaningless operation? Will nonlinear PCA do better at associating with specific regimes?

With the exception of competitive PCA, the algorithms presented in this chapter are based on single-step prediction for modeling the dynamics of a signal. However, from experimental results using feed-forward neural networks to model dynamical systems, we know that a small single-step prediction error variance is not necessarily a good indicator of whether the neural network has captured the longer term dynamics in the data. It would be interesting to explore the use of multistep prediction, where the output of a predictor is fed back to the input to form a closed loop, in either the training or testing phases of some of the algorithms. Furthermore, for chaotic signals there may be other more appropriate distance metrics than the experimental KL divergence, as measured through cross-prediction.

On a practical note, many of these algorithms require repeated training of the same neural networks. As a result, the symptoms of overtraining frequently appeared, such as unusually large network weights. Overtraining usually caused a degradation in segmentation performance and needs to be more seriously addressed than we have here.

We now summarize all the algorithms discussed at length in this chapter in Table 4.3.

4.10.2 Segmentation and Modeling in Practice

It is important to recognize that automatic segmentation and modeling of real-world data is a much more difficult task than we have eluded to here. In our opinion, variable scaling is the most serious problem affecting the performance of these algorithms on real data. It frequently happens that a signal is nonlinear at the source but is scaled through either the transmission or measurement process. A

Table 4.3 Comparison of Segmentation/Modeling Algorithms

Property Algorithm	Section Reference	(S)upervised (U)nsupervised	Likelihood Based (Y/N)	Temporal Scope: (S)equential (G)lobal	Assumed Switching Mechanism
CUSUM	4.2.2	S	Y	S	None
GLR	4.3	U	Y	S	None
SMOE → CUSUM	4.5.4	U → S	Y	G → S	None
MOE → CUSUM	4.5.1	U → S	Y	G → S	Function of time series
MPPCA → CUSUM	4.7.5	U → S	N	G → S	None
MOEPCA → CUSUM	4.7	U → S	N	G → S	Function of time series
ACE	4.8.2	U	Y	G	None
SOMCP	4.8.4	U	N	G	None
HCNN	4.9.2		N	G	None
MGE → Viterbi	4.9.4	U → S	Y	G → S	Markov chain
IOHMM → Viterbi MOC → Viterbi HMME → Viterbi	4.9.4	U → S	Y	G → S	Markov chain + function of time series

good example of this occurs in the collection of electroencephalogram (EEG) data (brain waves), which sometimes exhibits chaotic behavior. The conductivity between the electrodes and the scalp can change between and even within experiments. This can be fatal for some of the algorithms presented in this chapter because amplitude is important for nonlinear signals and a minor change in scaling can drastically affect the performance of a nonlinear predictor.

Real-world data may also be simultaneously subject to several different nonstationary changes, such as first-order changes in the mean superimposed on higher-order nonstationary changes. This can be due to the measurement process itself where sensors exhibit a bias drift over time. While we have emphasized completely unsupervised algorithms, any problem-specific prior information that is available should be incorporated and will likely lead to improved results. In addition, certainly not all changes are abrupt, and drifting between two regimes may actually be more common. However, useful results can often still be obtained by idealizing abrupt changes.

Real-world data may also be a switching among memoryless stochastic processes, nonmemoryless dissipative stochastic processes, and chaotic processes. However, if predictive models are used, the algorithms presented in this chapter will work for all nonmemoryless processes, even if there is switching between chaotic and nonchaotic processes. Therefore, in practice the data should first be probed to determine if it is memoryless. An open question, however, is how to handle switching among memoryless and nonmemoryless processes.

When segmenting an unknown signal in practice, we recommend using several algorithms independently and comparing the results, starting with ones that make few or no assumptions about the regime switching mechanism. These include the sequential GLR and global annealed techniques such as the ACE or SOMCP. The GLR provides sequential segmentation information, while the ACE and SOMCP can provide indication of recurring regimes.

In our opinion, not enough effort has been spent understanding the significance of the *production models* associated with these algorithms and the assumptions they represent. In particular, tests are needed to decide whether particular regime switching mechanisms are present in a particular time series. Recently, a few such tests have begun to appear in the economics literature. However, little is known as to whether there are certain classes of real systems that tend to exhibit particular types of switching. For example, do macroeconomic data tend to exhibit a different type of regime switching from geological data? Is the switching for currency exchange data any different than that typical of stock indices?

4.10.3 Final Words

Much research was originally devoted to the segmentation and modeling problem, particularly in the area of speech. However, it rapidly became apparent that any errors in the definitive assignments involved in segmentation created serious problems for any higher level processing. For this reason, probabilistic assignments of regime were preferred, and definitive assignments were postponed for later processing. Thus, HMMs became the dominant model for processing piecewise or quasi-stationary signals. Nevertheless, nonstationary signals dominate real-world signals and unsupervised segmentation, and modeling algorithms still have an important place in the arsenal of the engineer or scientist for determining when such signals are piecewise stationary. Unfortunately, the term *unsupervised* in this context is still somewhat of an oxymoron; considerable human oversight is still required. It is our hope that work in this area continue to progress.

List of Symbols

a	scalar
a	vector
A	matrix
$E[\bullet]$	expectation operator
$e(n)$	instantaneous error (residual or innovation)
g	gate output or mixing coefficients
H	hypothesis
J	cost function evaluated over entire data set
$J(n)$	instantaneous cost
K	Kullback–Leibler divergence or number of experts
l	log-likelihood ratio increment
L	likelihood or log-likelihood ratio
m,n	discrete-time indices
$N(\mu,\Sigma)$	Gaussian distribution with mean μ and covariance Σ
P	probability
p	probability density function (pdf)
R	autocorrelation matrix
$S(f)$	power spectrum as a function of the digital frequency f
T	time of abrupt change point between two regimes
u	eigenvector
U	matrix of eigenvectors
$\mathbf{U}_P$	matrix of first P principal eigenvectors
w	free (adaptable) parameter or a vector of free parameters
W	free (adaptable) matrix
$x(n)$	scalar-valued discrete-time series at time step n
$\mathbf{x}(n)$	vector-valued discrete-time series
$\mathbf{X}(n)$	vector or matrix of the entire past history of a time series
$\chi(n)$	vector of the most recent past of a time series (the output of a tapped-delay-line)
$\varepsilon(n)$	local mean square error estimate
η	learning rate
λ	eigenvalue
μ	mean

σ^2 variance
Σ covariance
θ parameter set
$\bullet^\dagger$ transpose of a vector or matrix
$\hat{\bullet}$ estimate or prediction of a value
$\|\bullet\|$ Euclidean vector norm

BIBLIOGRAPHY

Andersson, P., 1985, "Adaptive forgetting in recursive identification through multiple models," *Int. J. Control*, vol. 42, no. 5, pp. 1175–1193.

Andre-Obrecht, R., 1988, "A new statistical approach for the automatic segmentation of continuous speech signals," *IEEE Trans. Acoustics, Speech, Signal Processing*, vol. 36, no. 1, pp. 29–40.

Appel, U., and A. V. Brandt, 1982, "Adaptive sequential segmentation of piecewise stationary time series," *Inf. Sci.*, vol. 29, no. 1, pp. 27–56.

Armitage, P., 1947, "Sequential analysis with more than two alternative hypotheses and its relation to discriminant function analysis," *J. Roy Statist. Soc. Suppl.*, vol. 9, pp. 250–263.

Bar-Shalom, Y., and T. E. Fortmann, 1988, *Tracking and Data Association* (New York: Academic Press).

Basseville, M., 1988, "Detecting changes in signals and systems—a survey," *Automatica*, vol. 24, no. 3, pp. 309–326.

Basseville, M., and A. Benveniste, 1983, Sequential detection of abrupt changes in spectral characteristics of digital signals, *IEEE Trans. on Information Theory*, vol. IT-29, no. 5., pp. 709–724.

Basseville, M., and A. Benveniste, eds., 1986, *Detection of Abrupt Changes in Signals and Dynamical Systems* (Berlin: Springer).

Basseville, M., and I. V. Nikiforov, 1993, *Detection of Abrupt Changes, Theory and Application* (Englewood Cliffs, NJ: Prentice-Hall).

Baum, C. W., and V. V. Veeravalli, 1994, "A sequential procedure for multihypothesis testing," *IEEE Trans. Information Theory*, vol. 40, no. 6, pp. 1994–2007.

Baum, L. E., 1972, "An inequality and associated maximization in statistical estimation for probabilistic functions of markov processes," *Inequalities*, vol. 3, pp. 1–8.

Bengio, S., F. Fessant, and D. Collobert, 1996, "Use of modular architectures for time-series prediction," *Neural Proc. Lett.*, vol. 3, no. 2, pp. 101–106.

Bengio, Y., and D. Frasconi, 1996, "Input–output HMM's for sequence processing," *IEEE Trans. Neural Networks*, vol. 7, no. 5, pp. 1231–1249.

Bishop, C. M., 1995, *Neural Networks for Pattern Recognition* (Oxford: Clarendon Press).

Box, G. E. P., and G. M. Jenkins, 1976, *Time Series Analysis: Forecasting and Control*, rev. ed. (San Francisco: Holden-Day).

Brandt, A. V., 1983, "Detecting and estimating parameter jumps using ladder algorithms and likelihood ratio tests," *Proc. ICASSP*, pp. 1017–1020.

Brillinger, D., 1975, *Time Series: Data Analysis and Theory* (San Francisco: Holden-Day).

Broomhead, D. S., and G. P. King, 1987, "Extracting qualitative dynamics from experimental data," *Physica D*, vol. 20, nos. 2–3, pp. 217–236.

Cacciatore, T. W., and S. J. Nowlan, 1994, "Mixtures of controllers for jump linear and nonlinear plants," *Adv. Neural Information Proc. Syst.*, vol. 6, pp. 719–726.

Casdagli, M., S. Eubank, J. D. Farmer, and J. Gibson, 1991, "State space reconstruction in the presence of noise," *Physica D*, vol. 51, pp. 52–98.

Chu, C., 1995, "Time series segmentation: A sliding window approach," *Inform. Sci.*, vol. 85, pp. 147–173.

Dempster, A. P., N. M. Laird, and D. B. Rubin, 1977, "Maximum likelihood from incomplete data via the EM algorithm," *J Roy. Stat. Soc.*, Series B, vol. 39, pp. 1–38.

Deshayes, J., and D. Picard, 1986, "Off-line statistical analysis of change-point models using nonparametric and likelihood methods," in M. Basseville and A. Benveniste, eds., *Detection of Abrupt Changes in Sigrals and Dynamical Systems* (Berlin: Springer), pp. 103–168.

Diks, C., 1996, "Detecting differences between delay vector distributions," *Phys. Rev. E*, vol. 53, pp. 2169–2176.

Dony, R. D., and S. Haykin, 1995 "Optimally adaptive transform coding," *IEEE Trans. Image Proc.*, vol. 4, no. 10, pp. 1358–1370.

Fancourt, C., and J. Principe, 1996, "A neighborhood map of competing one step predictors for piecewise segmentation and identification of time series," *Proc. IEEE Int. Conf. Neural Networks*, vol. 4, pp. 1906–1911.

Fancourt, C., and J. Principe, 1997, "Temporal self-organization through competitive prediction," *Proc. ICASSP*, vol. 4, pp. 3325–3328.

Fancourt, C., and J. Principe, 1998, "Competitive principal component analysis for locally stationary time series," *IEEE Trans. Signal Processing*, vol. 46, no. 11, pp. 3068–3081.

Fancourt, C., and J. Principe, 2000, "On the use of neural networks in the generalized likelihood ratio test for detecting abrupt changes in signals," *Proc. IEEE-INNS-ENNS Int. Conf. Neural Networks*, vol. 2, pp. 243–248.

Fristch, J., M. Finke, and A. Waibel, 1997, "Adaptively growing hierarchical mixture of experts," *Adv. Neural Inform. Proc. Syst.*, vol. 9, pp. 459–465.

Fukunaga, K., 1990, *Statistical Pattern Recognition* (Boston: Academic Press).

Goldfeld, S. M., and R. E. Quandt, 1973, "A Markov model for switching regressions," *J. Econometrics*, vol. 1, pp. 3–16.

Hamilton, J. D., 1989, "A new approach to the economic-analysis of non-stationary time-series and the business-cycle," *Econometrica*, vol. 57, no. 2, pp. 357–384.

Hamilton, J. D., 1990, "Analysis of time-series subject to changes in regime," *J. Econometrics*, vol. 45, nos. 1–2, pp. 39–70.

Haykin, S., 1996, *Adaptive Filter Theory* (Upper Saddle River, NJ: Prentice-Hall).

Haykin, S., 1999, *Neural Networks: A Comprehensive Foundation* (Upper Saddle River, NJ: Prentice-Hall).

Hediger, T., A. Passamante, and M. E. Farrell, 1990, "Characterizing attractors using local intrinsic dimensions calculated by singular-value decomposition and information-theoretic criteria," *Phys. Rev. A*, vol. 41, no. 10, pp. 5325–5332.

Jacobs, R., F. Peng, and M. Tanner, 1997, "Bayesian approach to model selection in hierarchical mixtures-of-experts architectures," *Neural Networks*, vol. 10, no. 2, pp. 231–241.

Jacobs, R. A., M. I. Jordan, S. J. Nowlan, and G. E. Hinton, 1991, "Adaptive mixtures of local experts," *Neural Computation*, vol. 3, pp. 79–87.

Jolliffe, I., 1986, *Principal Component Analysis* (New York: Springer).

Jordan, M. I., and R. A. Jacobs, 1994, "Hierarchical mixtures of experts and the EM algorithm," *Neural Computation*, vol. 6, pp. 181–214.

Jordan, M. I., and L. Xu, 1993, "Convergence results for the EM approach to mixtures of experts architectures," M.I.T. Artificial Intelligence Lab, A.I. Memo No. 1458.

Juang, B. H., and L. R. Rabiner, 1985, "Mixture autoregressive hidden Markov models for speech signals," *IEEE Trans. Acoustics, Speech, Signal Processing*, vol. ASSP-33, no. 6, pp. 1404–1413.

Kadirkamanathan, V., 1995, "Recursive nonlinear identification using multiple model algorithm," *Proc. IEEE Workshop on Neural Networks for Signal Processing*, pp. 171–180.

Kalman, R. E., 1960, "A new approach to linear filtering and prediction problems," *Trans. ASME, J. Basic Engrg.*, vol. 82, pp. 34–45.

Kantz, H., and T. Schreiber, 1997, *Nonlinear Time Series Analysis* (Cambridge, UK: Cambridge University Press).

Kazakos, D., and P. Papantoni-Kazakos, 1990, *Detection and Estimation* (New York: Computer Science Press).

Kehagias, A., and V. Petridis, 1997a, "Predictive modular neural networks for time series classification," *Neural Networks*, vol. 10, no. 1, pp. 31–49.

Kehagias, A., and V. Petridis, 1997b, "Time series segmentation using predictive modular neural networks," *Neural Computation*, vol. 9, no. 8, pp. 1691–1709.

Kerestecioglu, F., 1993, *Change Detection and Input Design in Dynamical Systems* (Somerset, England: Research Studies Press).

Klein, J. L., 1997, *Statistical Visions in Time* (Cambridge, UK: Cambridge University Press).

Kohlmorgen, J., S. Lemm, K.-R. Muller, S. Liehr, and K. Pawelzik, 1999, "Fast change point detection in switching dynamics using a hidden Markov model of prediction experts," *Proc. Int. Conf. Artificial Neural Networks*.

Kohlmorgen, J., K.-R. Muller, and K. Pawelzik, 1997, "Segmentation and identification of drifting dynamical systems," *Proc. 1997 7th IEEE Workshop on Neural Networks for Signal Processing*, pp. 326–335.

Kohonen, T., 1982, "Self-organized formation of topologically correct feature maps," *Biol. Cybernetics*, vol. 43, pp. 59–69.

Kohonen, T., 1990, "The self-organizing map," *Proc. IEEE*, vol. 78, no. 9, pp. 1464–1480.

Kohonen, T., 1993, "Generalizations of the self-organizing map," *Proc. Int. Joint Conf. Neural Networks*, vol. 1, pp. 457–462.

Kramer, M. A., 1991, "Nonlinear principal component analysis using auto-associative neural networks," *AIChe J.*, vol. 37, no. 2, pp. 233–243.

Levin, E., 1993, "Hidden control neural architecture modeling of nonlinear time varying systems and its applications," *IEEE Trans. Neural Networks*," vol. 4, no. 1, pp. 109–116.

Liehr, S., K. Pawelzik, J. Kohlmorgen, S. Lemm, and K.-R. Muller, 1999, "Hidden Markov mixtures of experts for prediction of non-stationary dynamics," *Proc. IEEE Workshop Neural Networks for Signal Processing*, pp. 195–204.

Lorden, G., 1971, "Procedures for reacting to a change in distribution," *Annals of Mathematical Statistics*, vol. 42, no. 6, pp. 1897–1908.

Makhoul, J., 1981, "On the eigenvectors of symmetric Toeplitz matrices," *IEEE Trans. Acoustics, Speech, Signal Processing*, vol. ASSP-29, pp. 868–872.

Masnadi-Shirazi, M. A., and N. Ahmed, 1991, "Optimum Laguerre networks for a class of discrete-time systems," *IEEE Trans. Signal Processing*, vol. 39, no. 9, pp. 2104–2108.

Morrison, D., 1976, *Multivariate Statistical Methods* (New York: McGraw-Hill).

Narendra, K. S., and J. Balakrishnan, 1997, "Adaptive control using multiple models, *IEEE Trans. Automatic Control*, vol. 42, no. 2, pp. 171–187.

Nicolis, G., 1995, *Introduction to Nonlinear Science*, (Cambridge, UK: Cambridge University Press).

Niedzwiecki, M., 1990, "Identification of nonstationary stochastic systems using parallel estimation schemes," *IEEE Trans. Automatic Control*, vol. 35, no. 3, pp. 329–334.

Niedzwiecki, M., 1992, "Multiple model approach to finite memory adaptive filtering," *IEEE Trans. Signal Processing*, vol. 40, no. 2, pp. 470–473.

Niedzwiecki, M., 1994, "Identification of time-varying systems with abrupt parameter changes," *Automatica*, vol. 30, no. 3, pp. 447–459.

Nikiforov, I. V., 1995, "A generalized change detection problem," *IEEE Trans. Information Theory*, vol. 41, no. 1, pp. 171–187.

Page, E. S., 1955, "A test for a change in a parameter occurring at an unknown point," *Biometrika*, vol. 42, pp. 523–527.

Pawelzik, K., J. Kohlmorgen, and K. R. Muller, 1996, "Annealed competition of experts for a segmentation and classification of switching dynamics," *Neural Computation*, vol. 8, no. 2, pp. 340–356.

Petridis, V., and A. Kehagias, 1996, "Modular neural networks for MAP classification of time series and the partition algorithm," *IEEE Trans. Neural Networks*, vol. 7, no. 1, pp. 73–86.

Pinsker, M. S., 1964, *Information and Information Stability of Random Variables and Processes* (San Fransisco: Holden Day).

Plataniotis, K. N., D. Androutsos, A. N. Venetsanopoulos, and D. G. Lainiotis, 1996, "New time series classification approach," *Signal Processing*, vol. 54, no. 2, pp. 191–199.

Poritz, A. B., 1982, "Linear predictive hidden Markov models and the speech signal," *Proc. ICASSP*, pp. 1291–1294.

Priestlley, M. B., 1988, *Non-linear and Non-stationary Time Series Analysis* (San Diego: Academic Press).

Principe, J., B. de Vries, and P. Guedes de Oliveira, 1993, "The gamma filter: a new class of adaptive IIR filters with restricted feedback," *IEEE Trans. Signal Processing*, vol. 41, no. 2, pp. 649–656.

Quandt, R. E., 1972, "A new approach to estimating switching regressions," *J. Am. Stat. Assoc.*, vol. 67, pp. 306–310.

Quandt, R. E., and J. B. Ramsey, 1978, "Estimating mixtures of normal distributions and switching regressions," *J. Am. Stat. Assoc.*, vol. 73, no. 364, pp. 730–752.

Rabiner, L., and B. H. Juang, 1993, *Fundamentals of Speech Recognition* (Englewood Cliffs, NJ: Prentice-Hall).

Ramamurti, V., and J. Ghosh, 1999, "Structurally adaptive modular networks for nonstationary environments," *IEEE Trans. Neural Networks*, vol. 10, no. 1, pp. 152–160.

Ruanaidh, J. J. K., and W. J. Fitzgerald, 1996, *Numerical Bayesian Methods Applied to Signal Processing* (New York: Springer).

Sauer, T., 1992, "A noise reduction method for signals from nonlinear systems," *Physica D*, vol. 58, nos. 1–4, pp. 193–201.

Sauer, T., 1994, "Time series prediction using delay coordinate embedding," in A. S. Weigend and N. A. Gershenfeld eds., Time Series Prediction: Forecasting the Future and Understanding the Past (Reading, MA: Addison-Wesley), pp. 175–193.

Schreiber, T., 1997, "Detecting and analyzing nonstationarity in a time series with nonlinear cross predictions," *Phys. Rev. Lett.*, vol. 78, pp. 843–846.

Shi, S., and A. S. Weigend, 1997, "Taking time seriously: Hidden Markov experts applied to financial engineering," *Proc. 1997 IEEE/IAFE Conference on Computational Intelligence for Financial Engineering*, pp. 244–251.

Shiryaev, A.N., 1978, *Optimal Stopping Rules* (New York: Springer).

Srivastava, A. N., and A. S. Weigend, 1996, "Improved time series segmentation using gated experts with simmulated annealing," *Proc. 1996 IEEE Int. Conf. Neural Networks*, vol. 4, pp. 1883–1888.

Sugimoto, S., and T. Wada, 1988, "Spectral expression of information measures of Gaussian time series and their relation to AIC and CAT," *IEEE Trans. Information Theory*, vol. 34, no. 4, pp. 625–631.

Takens, F., 1980, "Detecting strange attractors in turbulence," in D. A. Rand and L. Young eds., *Dynamical Sytems and Turbulence* (Berlin: Springer), vol. 898, pp. 366–381.

Thomson, D. J., 1982, "Spectrum estimation and harmonic analysis," *Proc. IEEE*, vol. 70, no. 9, pp. 1055–1096.

Tipping, M. E., and C. M. Bishop, 1999, "Mixtures of probabilistic principal component analyzers," *Neural Computation*, vol. 11, no. 2, pp. 443–482.

Tong, H., and K. S. Lim, 1980, "Threshold autoregression, limit cycles, and cyclical data," *J. Roy. Stat. Soc. B*, vol. 42, pp. 245–292.

Van Trees, H., 1968, *Detection, Estimation and Modulation Theory: Part I* (New York: Wiley).

Verbout, S.M., J.M. Ooi, J.T. Ludwig, and A.V. Oppenheim," Parameter estimation for autoregressive Gaussian-Mixture Processes," *IEEE Trans. Signal Processing*, vol. 46, no. 10, pp. 2744–2756.

Waibel, A., T. Hanazawa, G. Hinton, K. Shikano, and K. J. Lang, 1989, "Phoneme recogntion using time-delay neural networks," *IEEE Trans. Acoustics, Speech, Signal Processing*, vol. 37, no. 3, pp. 328–339.

Wald, A., 1947, *Sequential Analysis* (New York: Wiley).

Waterhouse, S. R., and A. J. Robinson, 1995, "Non-linear prediction of acoustic vectors using hierarchichal mixtures of experts, *Adv. Neural Information Processing Syst.*, vol. 7, pp. 835–842.

Weigend, A. S., and N. A. Gershenfeld, eds., 1994, *Time Series Prediction: Forecasting the Future and Understanding the Past*, (Reading, MA: Addison-Wesley).

Weigend, A. S., M. Mangeas, and A. N. Srivastava, 1995, "Nonlinear gated experts for time series: Discovering regimes and avoiding overfitting," *Int. J. Neural Syst.*, vol. 6, no. 4, pp. 373–399.

Widrow, B., and S. D. Stearns, 1985, *Adaptive Signal Processing* (Englewood Cliffs, NJ: Prentice-Hall).

Wiener, N., 1960, *Extrapolation, Interpolation, and Smoothing of Stationary Time Series with Engineering Applications* (New York: Wiley).

Willsky, A. S., and H. L. Jones, 1976, "A generalized likelihood ratio approach to detection and estimation of jumps in linear systems," *IEEE Trans. Automatic Control*, vol. AC-21, no. 1, pp. 108–112.

Wold, H., 1938, *A Study in the Analysis of Stationary Time Series* (Uppsala, Sweden: Almquist and Wiksell).

Xu, L., 1994, "Signal segmentation by finite mixture model and EM algorithm," *Proc. Int. Symp. Artificial Neural Nets*, pp. 453–458.

Zeevi, A., R. Meir, and R. Adler, 1997, "Time series prediction using mixtures of experts," *Adv. Neural Information Processing Syst.*, vol. 9, pp. 309–315.

Zeevi, A., R. Meir, and R. Adler, 1999, "Non-linear models for time series using mixtures of autoregressive models," *J. Time Series Analysis*.

Zhang, X. J., 1989, *Auxiliary Signal Design in Fault Detection and Diagnosis*, lecture notes in control and information sciences, vol. 134 (Berlin: Springer).

5

APPLICATION OF FEEDFORWARD NETWORKS TO SPEECH

Shigeru Katagiri

5.1 INTRODUCTION

5.1.1 Why Are Feedforward Networks Applied to Speech?

A speech signal is an acoustic output of the human speech production system, which is the most fundamental communication medium filled with various types of information such as linguistic information, emotional information, and speaker identity. Therefore, a goal of speech processing technology has naturally been to extract and transmit such informaiton correctly and efficiently. For example, speech recognition has been aiming at accurately extracting linguistic information, such as word and sentence classes, conveyed in an utterance; speech coding has been aiming at transmitting, with a coded parameters of reduced sizes, information needed to maintain natural and easy-speech communications. In addition to the many traditional technological challenges, a recent powerful system selection, the artificial neural network (ANN), has been extensively applied to speech processing, aiming at a further improvement of speech processing technology in general.

There are many types of ANNs. Among them, a particular class of neural networks, feedforward networks (FFNs), has specially attracted great research interest in the speech processing area, because of their various promising characteristics. These characteristics include (1) a high data-modeling capability based on nonlinearity, (2) a high classification capability based on discriminative training, (3) a high trainability that reminds one of human learnability, and (4) a simple distributed mechanism suitable for high-speed computation. Actually, a typical FFN, called the multilayer perceptron (MLP) network has almost always played a central role in a wide range of ANN-based speech processing tasks, from speech recognition, speech coding, and speech enhancement, to speech synthesis. In addition, another class of FFNs called learning vector quantization (LVQ) (Kohonen 1995), which was developed with a close link to the self-organizing map (SOM) (Kohonen 1995) and can also be considered as a special case of the radial-basis function (RBF) network (e.g., Broomhead and Lowe 1998), has been extensively used to implement highly discriminative recognizers of speech signals.

As can be easily observed, a speech signal is a dynamic (temporal and nonstationary) and nonlinear signal. The FFN is indeed a nonlinear system and may be effective in modeling speech, but it is not necessarily fully appropriate for modeling the dynamics of speech, due to its simple feedforward architecture. One solution to this difficulty is to use a recurrent neural network (RNN) that represents the dynamics in its particular mechanism of information recurrency. This alternative choice of ANN naturally sounds quite promising, and actually, many studies on the application of RNNs to speech processing have been reported in the literature (e.g., Robinson 1994; Schuster and Paliwal 1997). However, in spite of the high potential of RNNs, a practical first-step solution still continues to be a hybrid system that incorporates a state transition mechanism with FNNs. The reasons for this system selection include (1) the reasonability and tractability of the state transition mechanism, which has been widely used in speech recognition techniques based on dynamic time warping (DTW) and the hidden Markov model (HMM) (e.g., Rabiner and Juang 1993), and (2) the insufficient capability of simple recurrency to model the complex dynamics of speech signals.

As cited above, the FFN is not a fully adequate system selection for speech processing. Nevertheless, at the same time, it is an obvious fact that the FFN constitutes an important subarea of ANN-based speech processing technology. Accordingly, speech processing using FFNs is clearly worth studying. In light of this, we dedicate this chapter to a comprehensive introduction of FFN applications to speech processing.

5.1.2 Chapter Organization

Discussions on FFN applications to speech processing require basic knowledge about the mechanism of human speech communications and the acoustic/linguistic nature of speech signals. In addition, for high readability, consideration should also be given to some consistent basis of problem formalization. Therefore, we provide brief summaries on speech communications and system design issues, though such topics have already been introduced in many other publications (e.g., Huang et al. 1990; Lee et al. 1996; Rabiner and Juang 1993).

There are actually various types of application topics, even though we limit our neural network selection to FFNs: speech recognition, speech coding, speech enhancement, speech synthesis, speaker recognition, and language acquisition. Among them, we focus on the topic of speech recognition, which has been most vigorously investigated.

This chapter is organized as follows. Section 5.2 summarizes the basic characteristics of speech signals and fundamentals of speech processing technologies. Section 5.3 overviews design issues of ANN systems, focusing on FFNs. Then, Section 5.4 introduces FFN applications to speech recognition, and Section 5.5 provides an archived introduction to other application topics, in particular, speech coding, speech enhancement, and speaker recognition. Finally, we conclude the chapter in Section 5.6.

5.2 FUNDAMENTALS OF SPEECH SIGNALS AND PROCESSING TECHNOLOGIES

5.2.1 Overview

Speech processing is a multidisciplinary research area that emcompasses various disciplines such as phonetics, linguistics, psychology, physiology, neurobiology, informatics, and computer engineering. This chapter is obviously insufficient for introducing even only the essence of the entire area. Having stated the obvious, we focus our brief introduction on a limited number of basic ideas and topics that are specific to speech signals and are crucial for understanding the speech processing technologies discussed in later sections.

Readers are guided to textbooks such as Denes and Pinson (1993), Fant (1973), Huang et al. (1990), Lee et al. (1996), and Rabiner and Juang (1993) for more detailed information about the human communication mechanism, the nature of speech sounds, and speech processing technologies.

5.2.2 Information Conveyed in Speech Signals

A fundamental scheme of human speech communications is effectively illustrated in the diagram of a speech chain shown in Fig. 5.1 (Denes and Pinson 1993). A speaker first encodes his/her intention to linguistic units such as words and phonemes, controls articulators such as the tongue and the lips through neural information paths, and then produces an acoustic speech signal. The speech sound produced is transmitted as air vibration. The sound arriving at the ears of listeners vibrates the eardrums and is then converted to its cor-

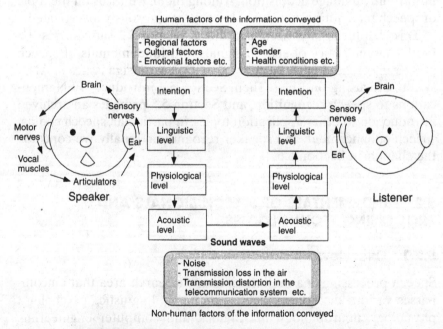

Figure 5.1 Diagram of the speech chain concept.

responding neural representation. This neural information travels up to the brain, it is then decoded to linguistic units, and ultimately it leads to the understanding of the speaker's intention.

As illustrated in Fig. 5.1, a speech sound conveys various kinds of interference information, such as the distortion in the electroacoustic and telecommunication system, as well as target information, such as linguistic information and the speaker's identity. Clearly, technology is expected to reduce such interference and to make it possible to transmit/recognize the target information effectively and accurately.

5.2.3   Humans Versus Speech Processing Systems

Humans live in a complex acoustic environment (auditory scene) where various sounds coexist, but they can still easily distinguish a target sound from others (Bregman 1990). This human ability is called the cocktail party effect, and it is widely recognized as a typical phenomenon that shows the unique abilities of humans. Indeed, one can easily become aware of the difficulty of separating/recognizing a target speech sound that is overlapped by other intereference sounds. Source signal separation is an ill-conditioned problem in which elements are determined only through the observation of the sum. Actually, compared with this human ability, the power of current (artificial) systems is seriously limited. What these systems aim to achieve is basically to code or recognize a speech signal that is *a priori* separated and segmented.

In fact, most recent research efforts have been made to improve the capability of coding/recognizing separated speech signals, although exceptional challenges are being made for ANNs to handle speech enhancement and blind separation. In much of the material presented in this chapter, we therefore assume the speech signals to be separated/segmented from other signals in advance and then provided to systems for further processing such as coding and recognition. Speech enhancement technologies are briefly introduced in Section 5.5.2.

5.2.4   Modular System Structure

Similar to most engineering systems, a speech processing system, such as a speech recognizer or a speech coder, employs a modular system structure. This system is composed of several modules, each executing a part of the entire process, that is, a subprocess. Among

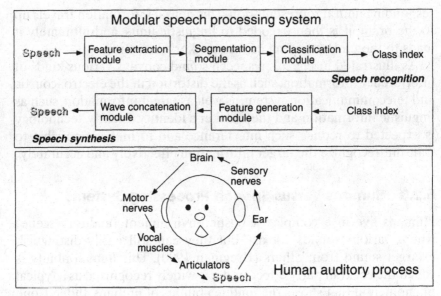

Figure 5.2 Schematic diagram of the human spoken process and a modular speech processing system.

many possibilities of modularization, a speech processing system typically consists of a front-end feature extraction module and a postend module used for coding or recognition.

Figure 5.2 illustrates the concept of modular speech processing systems and also illustrates an analogy between the human auditory process and its artifitical counterpart. Actually, encouraged by this analogy, modular system structures have been widely used in speech processing technologies.

5.2.5 Basic Feature Representations

Sound Spectrogram

Time–Frequency–Energy Representation A speech wave is a one-dimensional signal having a temporal structure. The signal can be considered a combination of different sine waves, and its acoustical characteristics are determined by the frequency, energy (amplitude), and phase of each component. In speech technology, however, a speech signal is usually represented as a three-dimensional, time–frequency–energy feature pattern that is similar to a sound spectrogram or a voice print: The three factors, that is, time, frequency, and energy, dominate the primary perceptual nature of basic linguistic units, that is, phonemes.

Figure 5.3 shows a sample sound spectrogram of the spoken word "Yokozuna" (the rank of grand champion wresler in Japanese sumo wrestling). In the spectrogram, a shaded point indicates the intensity of energy at a time–frequency point. The darker the shade is, the higher the intensity is. The gray-scaled sound spectrogram is basically a three-dimensional object with its height being the energy intensity. A section sliced at some time position in parallel to the frequency axis can therefore be observed as a two-dimensional frequency–energy pattern, that is, a short-time spectrum. The sound spectrogram is accordingly treated as a sequence of short-time spectra.

Nonlinear Time-Warping Structure As can easily be experienced, a speech sound varies in length, based on changes in the speaking speed (rate), though its linguistic information does not change. In addition, the temporal warping (extension and shrinking) range of speech is, in general, large for vowel classes and small for consonant classes. That is, the temporal structure of a speech sound is nonuniform, or in other words, nonlinearly time warped.

Obviously, it is not appropriate to model a nonlinearly warping, various-length sound spectrogram or its corresponding speech signal as a single fixed-length (fixed-dimensional) pattern. The desirable way of modeling should reflect the temporal structure specific to speech. A natural and widely used solution is to model the observed speech as a variable-length sequence of short-time spectra. In this scheme, the short-time spectra are usually modeled by using a fixed-dimensional vector format called an acoustic feature vector. Figure 5.4 illustrates speech signal modeling by a sequence of acoustic feature vectors.

Structure of a Short-Time Spectrum In a sound spectrogram, one can find that a short-time spectrum shows two noticeable structures: (1) spectral envelop and (2) harmonic or noisy structure. The spectral envelop is an outline structure that smoothly varies, and it is mainly determined by the status (shape) of articulators such as the lips and tongue. Linguistic information, which is the result of the articulators' behavior and is therefore important for speech recognition and transmission, is accordingly included mainly in the spectral envelop. The harmonic structure is the result of the vibration of

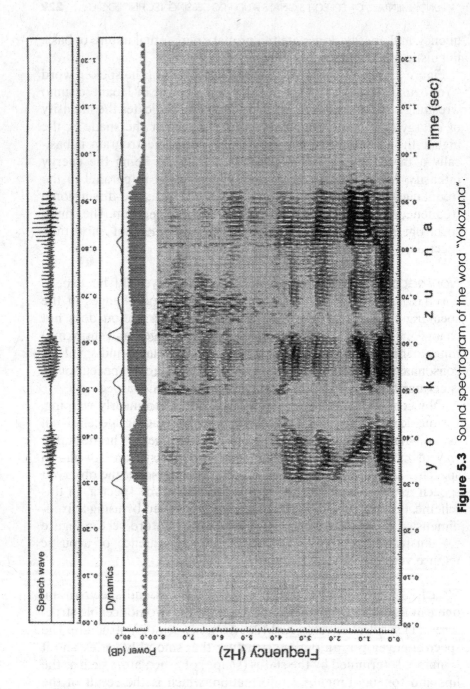

Figure 5.3 Sound spectrogram of the word "Yokozuna".

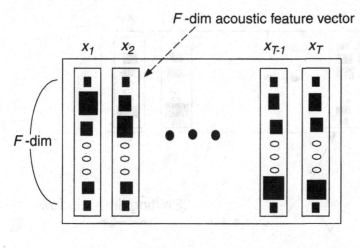

$$X_T = (x_1, x_2, \ldots x_{T-1}, x_T)$$

■ : Box size = Value of vector element

Figure 5.4 Speech signal modeling using a sequence of acoustic feature vectors.

the vocal cords; the noisy structure is caused by the air noise produced during the closure or constriction by the articulators. These latter structures are both based on the driving sources for speech production and are considered to mainly convey the speaker's identity and prosodic information.

Although driving-source-based structures apparently seem complicated, they can be modeled in a comparatively simple way. Basically, a harmonic structure can be modeled by a pulse train, and a noisy structure can be modeled by some statistical noise such as Gaussian noise. Based on the simplicity of such driving-source-based structures and the spectral envelop's richness of linguistic information, modeling the spectral envelop structure is a primary target of feature representation (modeling) for most speech signal processing technologies.

Acoustic Feature Vector   As cited above, the spectral envelop of every short-time spectrum is an important linguistic information source. Actually, it is known to be heavily redundant from the viewpoint of speech recognizer and coder design. Therefore, the acoustic feature vectors that are used for modeling such short-time spectra

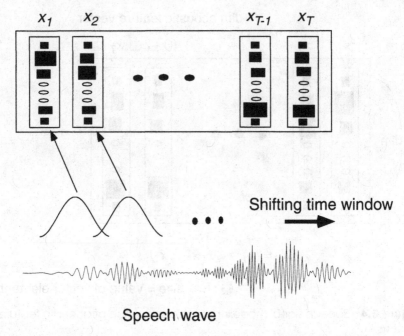

Figure 5.5 Mechanism for calculating acoustic feature vectors.

(and their corresponding spectral envelop shapes) are generally computed with some low vector dimensionality (usually at most 40) so that they can efficiently represent information needed for preset design goals.

For simplicity, every system design can choose to fix both the dimensionality and the temporal resolution of the acoustic feature vectors to some preset values; here, the temporal resolution means the step size of shifting a time window for computing every acoustic feature vector. Generally, based on experimental results of coding and recognition, acoustic feature vectors are calculated every 5 to 10 msec by shifting a time window with a length of 20 to 30 msec over the observed speech signal. The dimension of the acoustic feature vector is usually in a range from 10 to 40. Accordingly, a speech observation is generally modeled by a sequence of low-dimensional acoustic feature vectors, each of which is calculated every 5 to 10 msec by using a 20- to 30-msec time window. Figure 5.5 illustrates the mechanism of calculating the acoustic feature vectors.

There are several ways of calculating the acoustic feature vectors. In the following paragraphs, we introduce three methods widely used in present speech technologies. One should notice that each method is applied to every time window used for calculating its corresponding single acoustic feature vector.

Filter-Bank Method   The traditional way to compute the acoustic feature vectors is with a filter bank method. This method is based on the psychoacoustic finding that the human auditory system performs the function of filter-bank-like frequency analysis of input sounds.

A diagram of a filter bank method is illustrated in Fig. 5.6. A speech signal is passed through a bank of multiple bandpass filters. Generally, the individual filters are spaced according to a nonuniform frequency scale such as the Mel scale or Bark scale, based on perceptual experiments. Each filter outputs a kind of short-time energy that can be observed through its corresponding frequency range, and the collection of these outputs forms an acoustic feature vector. According to the limited number of bandpass filters and the smoothing effect of the individual filters, the resulting acoustic feature vector behaves as an effective estimate of the spectral envelope.

Autoregressive Modeling Method   Autoregressive modeling is a widely used parametric time-series modeling paradigm. In speech

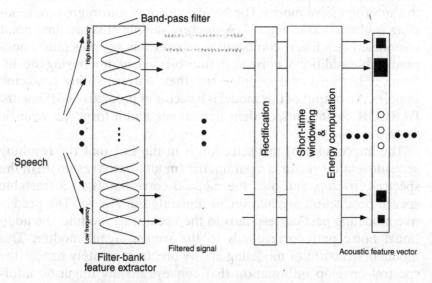

Figure 5.6 Diagram of the filter bank method.

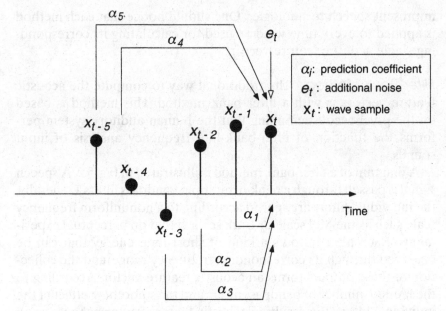

Figure 5.7 Concept of autoregressive modeling.

processing technologies, it is often referred to as the linear predictive coding (LPC) method or the PARCOR method, based on its prediction mechanism in the time domain or based on its efficient calculation mechanism using the partial correlation coefficients of the autoregressive model. The basic concept of autoregressive modeling is illustrated in Fig. 5.7. A discrete sample at some time point is modeled as a linear combination of previous samples plus a nonpredictable additional noise, with the objective of minimizing the differences between real samples and their corresponding predicted samples. An output of this model is a vector of predictive coefficients, PARCOR coefficients, or their derivatives, and it forms an acoustic feature vector.

The importance of this selection is in the fact that the resulting acoustic feature vector is a parametric (model-based) estimate of the spectral envelop, and also the method corresponds to a tractable speech production mechanism, as illustrated in Fig. 5.8. The predictive modeling part corresponds to the vocal tract module; the additional noise part corresponds to the source signal module. This modular structure of modeling allows one to separately handle the spectral envelop information that conveys mainly linguistic information and the harmonic or noise information that basically corre-

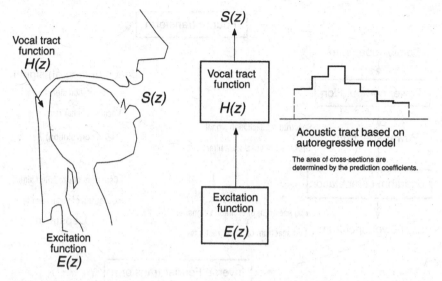

Figure 5.8 Vocal tract based on autoregressive modeling and an illustration of the human speech production system.

sponds to the speaker's identity and the prosodic information. The end result is an effective and efficient implementation of methods for speech coding, speech recognition, and speaker recognition.

Cepstrum Method Cepstrum is a type of homomorphic transform that is accomplished by a Fourier transform and a logarithmic operation, and it is defined as the inverse Fourier transform of the log power spectrum of a time-series signal. In the inverse Fourier transform stage, the frequency that is a variable of the spectrum function is remapped to a time-domain variable, specially called the quefrency. Figure 5.9 illustrates a diagram of the cepstrum computation.

Similar to a general spectrum analysis, the envelope structure of the spectrum (linguistic information) is represented in the low-quefrency region, and the minute structure of the spectrum (speaker's identity and prosodic information) is represented in the high-quefrency region. The separation of these two types of information is achieved by liftering, which is the filtering operation defined in the quefrency domain.

One may naturally notice that an appropriate lifter shape is needed for effective separation. In this light, various lifter shapes

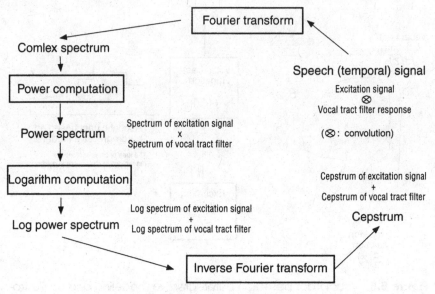

Figure 5.9 Diagram of cepstrum computation.

have actually been investigated (e.g., Ohyama et al. 1981). However, due to the lack of mathematical design criteria, the shape has usually been determined in a trial-and-error fashion. Unfortunately, such heuristically designed lifter shapes are not optimal.

Modeling of a Temporal Structure   As shown in Fig. 5.3, a speech signal can be approximately separated to somewhat stationary segments, each basically corresponding to phonemes. To model such a temporal structure effectively, the concept of the state transition mechanism, in which each state is considered to correspond to a stationary segment, has been widely used in the speech processing area.

In the speech processing area, there are two main frameworks of the state transition mechanism that are popular: HMM and the DTW method using a prototype matching concept. These two frameworks are closely linked to each other because of the tight mathematical link between the HMM-based probability measure and the DTW-based distance measure. In the following paragraphs, we briefly introduce the concept of the HMM, which appears repeatedly in later sections.

Figure 5.10 illustrates a sample left-to-right HMM that is considered suitable for the modeling of speech signals. The model consists of N_s states, each indicated by a circle. It is assumed that the transi-

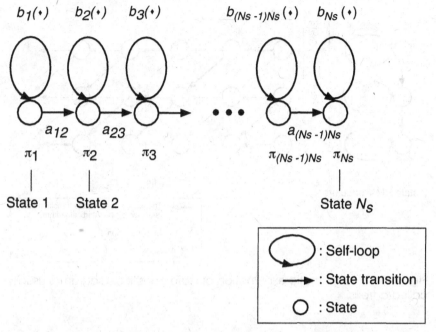

Figure 5.10 Structure example of a left-to-right HMM.

tion occurs in the probabilistic sense from one state to another or recursively within a single state according to the initial state probabilities $\{\pi_i\}$ ($1 \leq i \leq N_s$; $\pi = \{\pi_i\}$) and transition probabilities $\{a_{ij}\}$ ($1 \leq i,j \leq N_s$; $A = \{a_{ij}\}$). It should be emphasized here that this transition cannot be observed.

The model is also assumed to emit an observation $\mathbf{o}_t$, which is a vector in the acoustic feature vector space at hand, at observation time index t. The emission of this observation depends on the output probabilities $\{b_j(\cdot)\}$ [$b_j(\mathbf{o}_t)$ for observing $\mathbf{o}_t$ at state j; $B = \{b_j(\cdot)\}$]. Therefore, an observation sequence O_T is defined as the sequence of model outputs that is probabilisticly produced through the state transition and its corresponding observation emission; $O_T = [\mathbf{o}_1\mathbf{o}_2, \ldots, \mathbf{o}_t, \ldots, \mathbf{o}_T]$. The probability for observing O_T is accordingly provided as

$$P(O_T|\lambda) = \sum_{\text{all } \mathbf{q}} P(O_T|\mathbf{q}, \lambda) P(\mathbf{q}|\lambda)$$

$$= \sum_{q_1, q_2, \ldots, q_T} \pi_{q_1} b_{q_1}(\mathbf{o}_1) a_{q_1, q_2} b_{q_2}(\mathbf{o}_2) \times \cdots$$

$$\cdots \times a_{q_{T-1} q_T} b_{q_T}(\mathbf{o}_T) \tag{5.1}$$

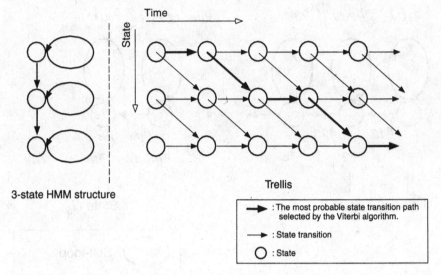

Figure 5.11 Schematic explanation of state transition; diagram is usually called a trellis.

where **q** is the state sequence represented as $\mathbf{q} = (q_1 q_2, \ldots, q_T)$, q_t is the state at t in **q**, and $\lambda = (A, B, \pi)$.

It should be noted here that the HMM probability (5.1) is the result of the probability accumulation over all possible state transition paths. A diagram of these possible state transition paths is referred to as a trellis and is illustrated in Fig. 5.11.

5.2.6 Speech Recognition Task Paradigms

Because of the large variety of speaking styles, various types of speech recognition task paradigms are possible, for example, isolated word recognition, read connected word recognition, and spontaneous (conversational) speech recognition. Generally, the recognition of spontaneous speech samples is more difficult than that of read speech samples: Often, spontaneous speaking does not involve the observation of grammar and also does contain speech segments that are not directly relevant to the tasks at hand, such as interjections and repairs.

An ultimate goal of speech recognition is to recognize all individual words or speech segments. To do this, a recognizer must represent each word (class) by its corresponding model, which is usually constituted by HMMs and/or ANNs. However, for large vocabulary

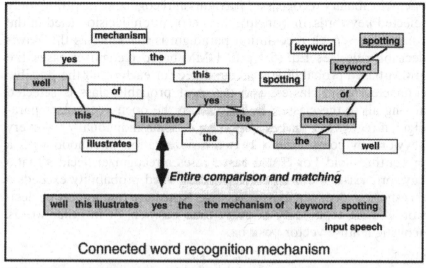

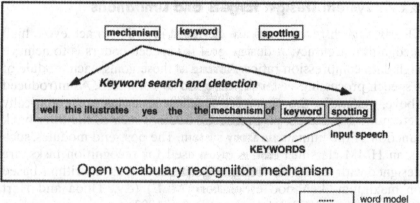

Figure 5.12 Mechanism for connected word recognition that aims to recognize all individual words and the mechanism for open-vocabulary recognition that aims to recognize only selected keywords.

tasks, it is quite difficult to implement all vocabulary models in a word-by-word mode. Generally, therefore, a word model is simply formed by the cocatenation of phoneme models, each comprising HMMs and/or ANNs. Moreover, an alternative to the recognition of the entire speech input, that is, an open-vocabulary recognition approach, is often employed.

Figure 5.12 illustrates the mechanism for connected word recognition (sometimes called continuous speech recognition) that aims at recognizing all individual words and the mechanism for open-vocabulary recognition that aims at recognizing (spotting) only selected keywords. In principle, the recognition decision used in the continuous speech recognition paradigm is the same as the Bayes decision rule [see Eq. (5.7)]. In HMM-based recognizer cases, the HMM-based probability is accumulated for each of all the possible connected word classes, and the most probable class is selected among all of the classes. In contrast, in the open-vocabulary paradigm, a recognizer makes a spotting decision individually for every keyword by comparing a keyword observation likelihood with a preset threshold. For HMM-based cases, a recognizer decides that a keyword exists when its corresponding word probability exceeds a threshold; it does not exist otherwise. Note here that a spotting decision is done repeatedly at every time position, or in other words, acoustic feature vector position.

5.2.7 System Design Targets and Limitations

Clearly, speech recognizers are designed in order to achieve a high recognition accuracy. A design goal for speech coders is to achieve high data compression ratios. Aiming at these goals, each module of a speech processing system is designed or optimized. As introduced above, generally, the feature extraction modules are empirically determined, based on psycoacoustic findings such as the filter bank function of the human auditory system. The post-end modules, such as an HMM classifier that is often used for recognition tasks, are designed with traditional design algorithms such as algorithms based on maximum-likelihood estimation (MLE) (e.g., Duda and Hart 1973; Fukunaga 1972; Rabiner and Juang 1993).

Typically, a different system is based on a different design approach that consists of a different type of acoustic feature vector, a different selection of the system structure (e.g., prototype matching or HMM), and a different selection of the design criterion [e.g., the MLE criterion or the maximum-mutual information (MMI) criterion]. Roughly speaking, in traditional design attempts at speech processing systems, the feature extraction modules are determined empirically, and the post-end modules are trained by using conventional design criteria such as the MLE criterion.

Such traditional systems have actually been shown to be useful in some limited conditions, but it has also been pointed out that they cannot perform satisfactorily in real-world large-scale applications. The most serious reason for this dissatisfaction is considered to be the limited optimization capability of the algorithms employed. These algorithms include a vector quantization (VQ) method (Gray 1984) based on the minimum distortion objective (e.g., k-means clustering) and MLE training for HMM speech pattern classifiers. Accordingly, expectations are high for emerging ANN-based technologies to alleviate the above dissatisfaction due to the traditional selection of optimization algorithms.

5.3 FUNDAMENTAL ISSUES OF ANN DESIGN

5.3.1 Overview

It was in the early stages of ANN applications to speech processing that the design algorithms motivated by ANN concepts attracted main research attention. For example, corrective training, which is motivated by the historic ANN training algorithm called error correction training (Nilsson 1990), was used in recognizer design to alleviate the limitations of the conventional MLE-based design algorithms (Bahl et al. 1988). After that, research concerns shifted to ANN applications of an advanced stage, and many speech processing systems came to be designed by using MLP, SOM, or LVQ networks (e.g., Kohonen 1995).

ANNs successfully demonstrated their effectiveness in various tasks in those early attempts. However, there was no clear notion of just how these ANNs related to improvements in performance. Moreover, the algorithms suffered from some mix-up of fundamental technical issues such as design objectives and optimization procedures, and consequently, their application results were not fully analyzed for further advancement.

The insufficient status or mix-up in understanding the characteristics of ANN design has been significantly improved since those days. Nowadays, it is clearly recognized that, similar to most examples of traditional system design, ANN designs should be attempted based on the following fundamental design issues:

1. Selection of ANN structures
2. Selection of training objectives
3. Selection of optimization algorithms
4. Selection of criteria for increasing the design robustness to unknown samples, which is sometimes referred to as the generalization capability

In the remaining of this section, we therefore review ANN design from the viewpoint of the fundamental issues just mentioned.

In particular, the issues have been reelaborated for recognizer design purposes in recent studies on the family of recognizer design methods, called the generalized probabilistic descent (GPD) method (e.g., Katagiri 1998). The formalization of GPD has been shown to be useful for discussing/analyzing speech recognizer design from a wide viewpoint encompassing ANN-related issues as well as the above-mentioned fundamentals. It is therefore befitting that we present a brief introduction to this emerging paradigm of recognizer design.

5.3.2 Structure

Overview   The selection of the structure apparently determines the ANN configuration, and it accordingly controls the information flow in the network. It basically controls how flexibly the network represents information (i.e., the representation capability of the network). To see mathematical discussions on the representation capability, the reader is referred, for example, to Chapter 2 of this book.

When we concentrate our discussions on FFNs, the main possibilities of structural selections involve only the two following types: (1) simple feedforward structures and (2) hybrids of FFNs and others.

Simple Feedforward Structures   Figure 5.13 illustrates the basic structure of a simple FFN. As illustrated in the figure, this FFN conveys node output signals in a one-way mode. The information goes up from the bottom layer of the network, which consists of input nodes, to the top layer. According to the size of the intermediate hidden layers (the number of hidden layer nodes), the network's bahaviors are classified into two distinctive groups:

1. Data compression based on a sandglass-shaped structure (the left side of Fig. 5.14)
2. Data regression or classification based on a lozenge-shaped structure (the right side of Fig. 5.14).

The former data compression function seemingly provides a vital framework for speech coding and enhancement. The information

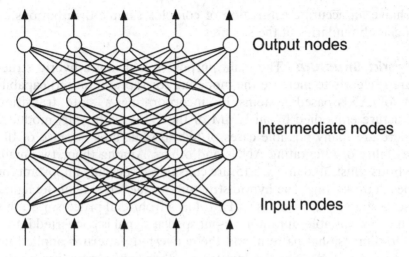

Figure 5.13 Structure of a feedforward neural network.

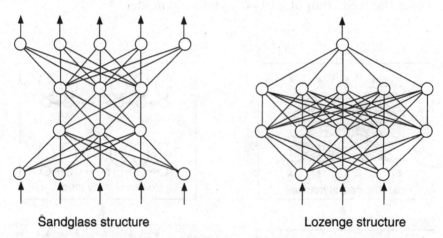

Sandglass structure   Lozenge structure

Figure 5.14 Two types of feedforward neural network structures: Sandglass structure for data compression (left) and lozenge structure for data regression or classification (right).

represented at the constricted intermediate layers of the network is expected to represent the principal information of speech signals useful for data transmission and storage.

The latter regression or classification function plays a central role in designing speech recognizers. The information spanned at the high-dimensional intermediate layers of the network enables one to achieve an accurate estimation of complex sample distributions or of class boundaries of the samples.

Hybrid Structure   The main purpose for having a hybrid structure is clearly to increase the modeling (or representation) capability of ANN-based systems by incorporating a state transition structure embodied by an HMM or the DTW with ANN modules. There are many possible cases of hybrid structures because of the flexibility of integrating ANNs and others. Among them, two main hybrids, illustrated in Fig. 5.15, are considered specially important for speech processing. The hybrid structure on the left side of the figure uses a state model such as an HMM as a front-end processor. In this scheme, a variable-durational input speech signal is converted into a fixed-dimensional pattern, and the converted pattern is applied up to a post-end ANN as the input. The other hybrid structure on the right side of the figure first uses an ANN at the front-end stage and passes the ANN output to a post-end state model.

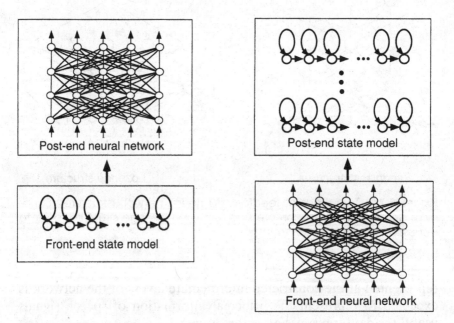

Figure 5.15 Two types of neural-network-based hybrid systems: Hybrid system using a front-end state model and a post-end neural network (left) and hybrid system using a front-end neural network and a post-end state model (right).

A natural design principle is to optimize the entire hybrid system globally and consistently. Conventionally, however, the components (modules) of hybrid systems are independently designed, and the resulting components are simply integrated in a hybrid structure.

5.3.3 Measurement

Overview

Neural Node Functions Determining Measurements The measurement selection determines how the network expresses the information contained in the input signal. It is mainly characterized by the selections of two types of neural node functions: (1) basis functions and (2) activation functions. Figure 5.16 illustrates a general structure of a neural node that consists of these functions and is associated with K connections from the adjacent low layer.

The most popular selection for measurement in the current signal processing technology is "probability." In particular, probability has formed a central mathematical framework for speech recognition. In ANN formalisms too, the problem of estimating *a posteriori*

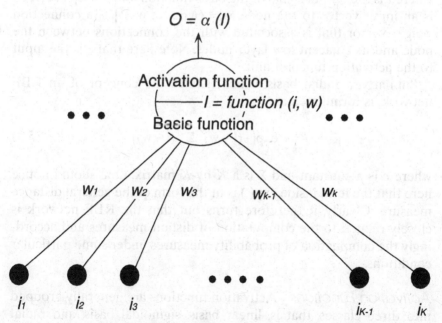

Figure 5.16 Structure of a neural node.

probabilities, which play an important role in recognition decisions, has been vigorously investigated (e.g., Gish 1990; Richard and Lippmann 1991).

On the other hand, ANNs using distance measures, such as SOMs and LVQ systems, have also constituted a major subarea of network research. Such measures, however, can have other more general interpretations than either probability or distance, for example, possibility. This kind of interpretation is useful for discussing designs of ANN pattern classifiers from the perspective of the traditional Bayes decision theory.

Basis Functions   There are two main types of selections for the basis function: (1) linear basis and (2) radial basis. For explanation purposes, let us consider the neural node illustrated in Fig. 5.16.

A linear basis forms a perceptron-type network such as MLP, and it is basically formulated as

$$I = \mathbf{i} \cdot \mathbf{w} = \sum_{k=1}^{K} w_k i_k \tag{5.2}$$

where I is the output signal of the basis function unit, $\mathbf{i}[= (i_1, \ldots, i_K)^T]$ is an input vector to the node, $\mathbf{w}[= (w_1, \ldots, w_K)^T]$ is a connection weight vector that is associated with the connections between the node and its adjacent low layer nodes. Note here that I is the input to the activation function unit.

Similarly, a radial basis, which defines the concept of an RBF network, is formulated as

$$I = c \exp\left(-(\mathbf{i} - \mathbf{w})^T \Sigma^{-1} (\mathbf{i} - \mathbf{w})\right) \tag{5.3}$$

where c is a constant and Σ is a K-by-K matrix. One should notice here that the term inside exp( ) is of the form of the general distance measure. Clearly, it therefore turns out that the RBF network is closely related to the computation of distant measures and accordingly the compuation of probability measures under some particular conditions.

Activation Functions   Activation functions are generally grouped into three classes, that is, linear basis, sigmoidal basis, and radial basis.

A linear basis activation is simply a linear mapping, defined as

$$O = \alpha(I) = aI + b \tag{5.4}$$

where O is the output of the activation function, which is equivalent to the output of the corresponding node, and a and b are constants. In the case of a linear activation function, the measurement computed at a neural node is a straightforward linear function of the basis function I. When I is of the radial-basis form of (5.3), the output of the linear activation function resembles a distance or probability measure. The classic single-layer perceptron applies this linear activation function to the output of the linear basis function.

A sigmoidal basis activation is formulated as

$$O = \alpha(I) = \frac{1}{1 + \exp(-\alpha I + b)} \tag{5.5}$$

where a and b are constants, and a radial-basis activation is formulated in the same way as in (5.3). The sigmoidal activation, which has a smoothed version of the historic logistic function used in the classic ANN formalism, is a crucial component of recent multilayered FFNs such as MLP. Among these three types, the sigmoidal basis has been the most widely used.

Typical Combinations of Basis and Activation Functions

The combination of the basis and activation functions determines, in terms of the measurement, the type of FFN. There are two typical, widely used FFN types: perceptron-type network and RBF network.

Perceptron-Type Network   One of the most popular FFNs, the perceptron-type network, is defined by employing a linear basis function (5.2) and is also characterized by a sigmoidal activation function (5.5) used at the intermediate layer nodes.

A principal function of this type of network is the nonlinear mapping of an input pattern to a hyperplane through the linear basis and sigmoidal activation computations. The mapping is executed at every layer, and it emulates some feature extraction process. Because of this computation mechanism, a layered perceptron-type network is considered to have a larger capability of feature extraction. However, a layered network is often difficult to train (optimize) in reality; a three-layer (one hidden layer) network has been shown to have a significant capability of function approximation [e.g., see Funahashi (1989) for mathematical aspects]. Actually, three- or four-layer networks are widely used in most speech processing applications.

Radial-Basis Function Network   The RBF network is obtained by using a radial-basis function (5.3), and it usually uses a sigmoidal activation function at intermediate layer nodes. Similar to MLP, one can form a multiple-layered RBF network. However, probably in light of the relation between the RBF computation and traditional prototype-based distance computations, a simple three-layer (one hidden layer) RBF network is generally used, where the radial-basis functions are computed only at the hidden layer nodes.

In (5.3), treating the network connection vector **w** as a prototype vector or the average vector of a Gaussin distribution model, one can easily find that the measurement used by the RBF network is closely linked to distance measures such as the Euclidean distance and likelihood measures based on the Gaussian distribution. Focusing on this similarity, the RBF network has been used, with some additional conditions, as a likelihood network and a distance network in some applications (Katagiri et al. 1991a). This interpretation in terms of measurement is actually useful for integrating other systems such as HMMs to a hybrid system; for example, the likelihood network helps in the use of an HMM-based likelihood and the distance network helps in the use of a prototype-based distance.

5.3.4 Objective Function

Overview   Even if an ANN has a significant representation capability, it must be trained, adapted, or designed appropriately just like the human brain must be trained through education and experience. The design of an ANN is usually set by selecting an objective function, which is also often called a loss function, a risk function, and a criterion.

A popular objective is the squared error, which is computed between the output signal of the network and its corresponding target signal. This is widely used for data regression, data coding, and even for pattern classification. Another selection may involve an objective function that closely emulates the classification error count. This is one of the emerging concepts in the ANN and pattern recognition fields, and will be highlighted later on in the introduction to GPD.

In the three successive subsections, we shall summarize three main objective functions: the squared error objective, the classification error objective, and the mutual information objective. Basically, an observed speech pattern is of the vector sequence form as in (5.1), and accordingly one needs some special settings of objective functions that effectively reflect the dynamics of the speech input and task goal. However, for simplicity, we shall assume that for a given pattern X, an ANN-based system simply emits an M-dimensional output vector $\mathbf{o}$ [$= (o_1, \ldots, o_M)$]. Here, X can be either a fixed- or variable-length pattern, and M is a constant that expresses either the number of samples to be modeled in regression tasks or the number of possible classes in classification tasks.

Squared Error Objective We assume that for a given pattern X and its corresponding output vector $\mathbf{o}$, an M-dimensional target vector (teaching vector) $\mathbf{t}$ [$= (t_1, \ldots, t_M)$] is given to optimize the network. The assumed situation is illustrated in Fig. 5.17. Then, the squared error objective is defined as

$$\ell(X; \Lambda) = \|\mathbf{o} - \mathbf{t}\|^2 \qquad (5.6)$$

where $\|\ \|$ is the Euclidean distance between the two enclosed vectors, and Λ is a set of optimizable parameters of the network, such as network connection weights.

A design target here is to reduce the squared error objective over possible design data sets. Clearly, reducing the squared error objective will lead to a good approximation of the target vector. This approximation capability is indeed straightforward for an ANN-based reproduction of a speech signal, and therefore, this objective is a central selection used for speech coding and speech enhancement applications.

Classification Error Objective In classification problems, a design goal is not to approximate some target signal such as the teaching vector and to reduce classification errors (increase the classification accuracy). Actually, in the Bayes' decision theory, the minimization of error counts is shown to lead to the desirable, minimum classification error probability condition as shown here:

$$C(X) = C_i, \qquad \text{iff } i = \arg \max_{j=1,\ldots,M} g_j(X; \Lambda) \qquad (5.7)$$

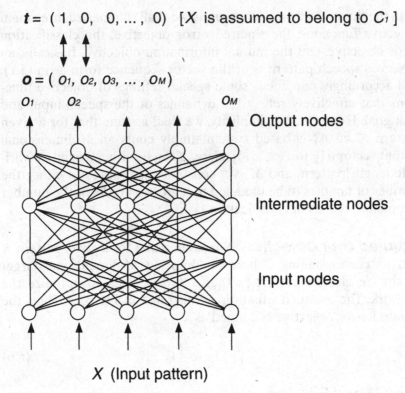

$t = (1, 0, 0, \ldots, 0)$ [X is assumed to belong to C_1]

$o = (o_1, o_2, o_3, \ldots, o_M)$

$o_1 \quad o_2 \qquad\qquad o_M$

Output nodes

Intermediate nodes

Input nodes

X (Input pattern)

Figure 5.17 Schematic explanation on the use of the squared error objective.

where X is the input pattern, $C(\cdot)$ is a classification operation (made by a classification system such as an ANN), C_j is the jth class index, M is the number of possible classes, Λ is a set of designable system parameters, and $g_j(X; \Lambda)$ is a discriminant function for C_j that indicates the degree (measured with Λ) to which X belongs to C_j. Figure 5.18 illustrates a scheme of the classification decision rule (5.7).

In the case of implementing the classification process of (5.7) with an ANN, discriminant functions, $g_j(X; \Lambda)$ ($j = 1, \ldots, M$), correspond to the elements of network output vector $\mathbf{o}$, that is, o_j's. The natural intuition here is that one first embeds $g_j(X; \Lambda)$'s or $\mathbf{o}$ in some objective that reflects a classification result (correct classification or misclassification) and optimizes Λ over possible design data so as to increase the classification accuracy.

The classification error objective is a direct implementation of the error count, and a typical example of it is provided as follows; for an input pattern X of C_k,

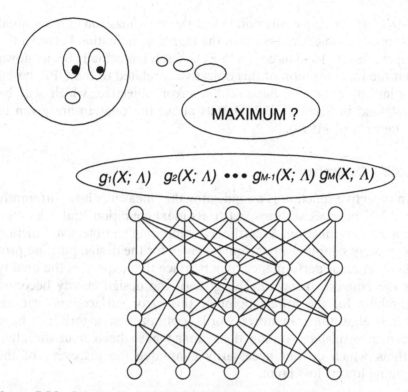

Figure 5.18 Schematic explanation on the use of the Bayes' decision rule for classification.

$$\ell_k(C(X)) = \begin{cases} 0 & (C(X) = C_k) \\ 1 & (\text{otherwise}) \end{cases} \tag{5.8}$$

Mutual Information Objective A design concept underlying the use of the mutual information objective is to increase the inter-class distinction by maximizing the mutual information between a sample belonging to some class and other classes. The mutual information is defined, for a sample X of C_k, by using the conditional probability function as follows:

$$I_k(X;\Lambda) = \ln \frac{p_\Lambda(X|C_k)}{\sum_j^M p_\Lambda(X|C_j)P(C_j)} \tag{5.9}$$

Some design or training is done over the system parameter Λ, aiming at increasing the value of this objective.

In (5.9), one can easily notice that the maximization of the mutual information objective results in the high discrimination between the target class (C_k) and its competing classes. It has actually been known that the maximization of this objective is related to the GPD-based minimization of the classification error objective, which will be explained in Section 5.3.7. Details about the relation are given in Katagiri et al. (1998).

5.3.5 Optimization

An objective function is an indicator that measures how differently an ANN behaves with respect to its ideal condition, and it is often called a loss function. Accordingly, computing an objective function over many samples (input signals) given for the design purpose produces an error performance hypersurface that expresses the quality of the network behavior. The goal of the design clearly becomes searching for the lowest point of this error surface, and various search algorithms, in other words, optimization algorithms, have been investigated. Among them, many have been heuristic algorithms, which do not necessarily guarantee the discovery of the optimal lowest loss point.

Local Search Versus Global Search   Mathematically supported algorithms are usually grouped in three classes: (1) gradient search (e.g., Fu 1968), (2) simulated annealing (e.g., Geman and Geman 1984), and (3) genetic algorithms (e.g., Holland 1992). Gradient search algorithms are traditional optimization algorithms that use the gradient of the error surface to find its lower positions. Algorithms of this type are fast and easy to handle, but they only guarantee the finding of local optimal points, which are not necessarily the lowest points of the error surface. Based on their high practicality, however, they are used most widely in the design of ANNs. Most of the ANNs used for speech processing actually use search algorithms of this type.

Simulated annealing algorithms were originally motivated by physical systems and have a significant feature that enables them to achieve the global optimal (minimum error) point of a surface in the probabilistic sense.

Genetic algorithms are based on biological evolution, and they aim at the global optimal point of a surface by repeating gene recombination operations. These global optimization algorithms often require a long training time and are not therefore popular in applications to speech processing.

Batch Optimization Versus Adaptive Optimization In the above paragraphs, we have summarized optimization methods from the viewpoint of the locality of the achievable optimality. There is actually another point of view in discussing optimization methods, that is, batch optimization versus adaptive (sequential) optimization. Batch optimization is a traditional scheme that attempts to find the optimal points of an error surface computed over all of the possible design samples (samples provided for training/design). The steepest descent method is a typical example of this scheme. In contrast, adaptive optimization aims at the optimal points of a surface every time a design sample is provided. The error back-propagation method, which plays a key role in demonstrating the utility of MLP networks, is a well-known example of the adaptive scheme (e.g., Rumelhart et al. 1986). Originally, ANN design was strongly motivated by the human ability of learning. Accordingly, adaptive optimization schemes, which do remind one of the high flexibility and adaptivity found in human learning, have attracted a lot of research attention.

5.3.6 Robustness to Unknown Samples

System design is unavoidably done using only limited design resources such as design samples and system parameters. It is naturally quite difficult for such real design attempts to cope with unknown future samples perfectly.

Indeed, many methods and approaches have been investigated to alleviate this problem, which is probably the most difficult issue in the general framework of system design and has also long formed a mainstream of research. To date, however, only few mathematical methods are viable for application-oriented, large-scale design attempts; actually in speech proceesing applications, several heuristic and empirical methods, for example, a method of increasing the smoothness of an ANN mapping function, have been tested (e.g., Katagiri 1994).

5.3.7 GPD-Based Design

Formalization

Introduction Generalized probabistic descent is a family of pattern recognizer design methods, which can be applied to various types of system structures and task settings. The significance of its development is to provide a novel mathematical perspective of discriminant function design as well as practical powerful solutions to the design (Juang and Katagiri 1992a,b; Katagiri et al. 1991a,b, 1994, 1998; Katagiri 1994).

For a description of GPD, let us consider the M class pattern classification task defined in (5.7). We assume our classifier's designable' parameters to be Λ and aim to find a status of Λ that corresponds to the minimum misclassification count condition. Our system measures the possibility with which an input X belongs to C_j by using the discriminant function $g_j(X; \Lambda)$. The larger $g_j(X; \Lambda)$ is, the higher the possibility X belongs to C_j.

In principle, any classifier design method should learn the nature of samples to be classified and reflect it to the status of the designable parameters. In the GPD formalism, a classifier is usually assumed to adjust its parameters (Λ in our case) every time a sample is provided for training (learning), aiming at the desirable minimum misclassification count condition.

Functional Form Embodiment of the Classification Process
The fundamental concept of the GPD formalization is to directly embed the overall process of classifying an input pattern in a smooth functional form (at least first-order differentiable with respect to the classifier parameters) that is suited for the use of a practical optimization method (Juang and Katagiri 1992b; Katagiri et al. 1991b). Based on a careful observation of the decision procedure in (5.7), GPD first defines a misclassification measure that emulates the procedure in a tractable smooth functional form. Among many possibilities, the following is a typical definition for a design sample X ($\in C_k$):

$$d_k(X;\Lambda) = -g_k(X;\Lambda) + \left[\frac{1}{M-1} \sum_{j, j \neq k} \{g_j(X;\Lambda)\}^{\mu} \right]^{1/\mu} \quad (5.10)$$

where μ is a positive constant. Clearly, we notice here that $d_k(\ ) > 0$ indicates a misclassification and $d_k(\ ) < 0$ indicates a correct classification. That is, the classification decision result of (5.7) is expressed by a simple scalar value. In addition, we find that controlling μ enables the simulation of various decision rules. In particular, when

μ approaches ∞, (5.10) resembles rule (5.7). An additional point of importance in terms of this measure is that the misclassification measure defined using the probabilistic discriminant functions is a kind of mutual information objective, which is explained in Section 5.3.4.

The next step of the GPD formalization is to introduce an objective function for every design sample such as X for evaluating classification decision results over design samples and specially define the function as a smooth, monotonically increasing function of the misclassification measure:

$$\ell_k(X;\Lambda) = l(d_k(X;\Lambda)) \qquad (5.11)$$

where $l(\cdot)$ is a scalar function that determines the characteristics of the objective function. Similar to the misclassification measure case, there are many possibilities for defining the objective. Among them, the GPD method usually uses the following smooth sigmoidal function:

$$\ell_k(X;\Lambda) = l(d_k(X;\Lambda)) = \frac{1}{1+e^{-(\alpha d_k(X;\Lambda)+\beta)}} \quad (\alpha > 0) \qquad (5.12)$$

where α and β are constants. Note here that the objective of (5.12) is a smoothed version of the error count objective of (5.8).

Optimization Based on the Probabilistic Descent Theorem
Following the Bayes' decision theory, the optimal status of Λ, which corresponds to the minimum misclassification probability condition, should be searched for by using the expected loss consisting of the objective function defined in (5.12), defined over all of the possible design samples as follows:

$$L(\Lambda) = \sum_k \int_\Omega p(X,C_k)\ell_k(X;\Lambda)1(X \in C_k)\,dX \qquad (5.13)$$

where Ω is the entire sample space of the patterns X's; it is assumed that $dp(X, C_k) = p(X, C_k)\,dX$, and that $1(\cdot)$ is an indicator function. The GPD's parameter optimization uses the adaptive adjustment training scheme based on the following probabilistic descent theorem for the expected loss (Amari 1967).

Probabilistic Descent Theorem   Assume that a given design sample X belongs to C_k. If the classifier parameter adjustment $\delta\Lambda(X, C_k, \Lambda)$ is specified by

$$\delta\Lambda(X,C_k,\Lambda) = -\varepsilon \mathbf{U} \nabla \ell_k(X;\Lambda) \tag{5.14}$$

where $\mathbf{U}$ is a positive-definite matrix and ε is a small positive real number, which is called a learning weight, then

$$E[\delta L(\Lambda)] \leq 0 \tag{5.15}$$

Furthermore, if an infinite sequence of randomly selected samples $X(t)$ is used for the design [$X(t)$ is a design pattern sample given at time index t in the design stage] and the adjustment rule of (5.14) is utilized with a corresponding learning weight sequence $\varepsilon(t)$, which satisfies

$$\sum_{t=1}^{\infty}\varepsilon(t) \to \infty \quad \text{and} \quad \sum_{t=1}^{\infty}\varepsilon(t)^2 < \infty \tag{5.16}$$

then the parameter sequence $\Lambda(t)$ (i.e., the state of Λ at t) according to

$$\Lambda(t+1) = \Lambda(t) + \delta\Lambda(X,C_k,\Lambda(t)) \tag{5.17}$$

converges with probability one (1) at least to Λ^*, which results in a local minimum of $L(\Lambda)$.

In a realistic situation where only finite design samples are available, the state of Λ that can be achieved by probabilistic descent training is at most a local optimum over a set of design samples. Morever, in the case of finite learning repetitions, Λ does not necessarily achieve even the local optimum solution. However, based on the directness of the formalization that embeds the classification process in an optimizable functional form, the high utility of GPD has been clearly demonstrated in various speech recognition tasks (e.g., Katagiri et al. 1998).

Minimum Classification Error Learning

Fundamental Concept   As repeated several times, the ultimate goal of recognizer/classifier design is to find the parameter set that achieves the minimum error condition. We summarize in this sub-section a theoretical advancement of the GPD formalization, which shows that a GPD-based minimization of the expected smooth classification error count loss possesses a fundamental capability for achieving this minimum error condition.

Let us assume that a probability measure $p(X)$ is provided for a pattern sample X. We first consider an unrealistic case where the true functional form of the probability is known. We also assume that a parameter set determining the functional form is $\dot{\Lambda}$, and consider a discriminant function that is computed as follows:

$$g_j(X;\dot{\Lambda}) = p_{\dot{\Lambda}}(C_j|X) \tag{5.18}$$

The classification rule used here is (5.7). Then, for example, using the misclassification measure of (5.10), we can rewrite the expected loss, defined by using the smooth classification error count loss (5.12) in the GPD formalization, as follows:

$$L(\dot{\Lambda}) = \sum_k \int_\Omega p(X,C_k)\ell_k(X;\dot{\Lambda})1(X \in C_k)\,dX$$
$$\simeq \sum_k \int_\Omega p(X,C_k)1(X \in C_k)$$
$$1\Big(p_{\dot{\Lambda}}(C_K|X) \neq \max_j p_{\dot{\Lambda}}(C_j|X)\Big)\,dX \tag{5.19}$$

Controlling the smoothness of functions such as the L_p norm used in (5.10) and the sigmoidal function used in (5.12), we can arbitrarily make $L(\dot{\Lambda})$ closer to the last equation in (5.19). Note here that we use $\dot{\Lambda}$. Based on this fact, the status of $\dot{\Lambda}$ that corresponds to the minimum of $L(\dot{\Lambda})$ in (5.19) (which is achieved by adjusting $\dot{\Lambda}$) is clearly equal to the $\dot{\Lambda}^*$ that corresponds to a true probability, or in other words, achieves the maximum *a posteriori* probability condition. Accordingly, the minimum condition of $L(\dot{\Lambda})$ can become arbitrarily close to the minimum classification error probability

$$\mathcal{E} = \sum_k \int_{\Omega_k} p_{\dot{\Lambda}^*}(X,C_k)1(X \in C_k)\,dX \tag{5.20}$$

where Ω_k is a partial space of Ω that causes a classification error according to the maximum *a posteriori* probability rule, that is,

$$\Omega_k = \left\{ X \in \Omega \;\middle|\; p_{\hat{\Lambda}*}(C_k|X) \neq \max_j p_{\hat{\Lambda}*}(C_j|X) \right\} \qquad (5.21)$$

Practical Utility The above result is quite interesting, since it shows that one can achieve the minimum classification error probability situation without the estimation of the *a posteriori* probabilities of all classes. However, the assumption that Λ is known is obviously unrealistic. Recent results concerning the approximation capability of ANNs have provided useful suggestions for studying this inadequacy. In the literature (e.g., Funahashi 1989), it has been shown that a three-layer perceptron has the fundamental capability of approximating an arbitrary function. In Hartman et al. (1990) and Park and Sandberg (1991), additionally, it is shown that the RBF network has a universal approximation capability. Furthermore, in Sorenson and Alspach (1971), a mixtured Gaussian distribution function is shown to approximate an arbitrary function in the sense of the minimum L_p norm distortion. In Chapter 2, one ongoing study on the approximation capability is introduced. Based on these results, one can argue that an ANN classifier with sufficient adjustable parameters has the fundamental capability of modeling an (unknown) true probability function. If $L(\Lambda)$ (and $\mathcal{E}$) has a unique minimum, and if Λ and $L(\Lambda)$ (and $\mathcal{E}$) are monotonic to each other, then the minimum corresponds to the case in which $g_j(X; \Lambda)$ is equal to a true probability function. It therefore follows that under this assumption, which is softer (more realistic) than that in the preceding paragraph, the GPD-based design enables one to fundamentally achieve the minimum classification error probability, even if the true parametric form of the probability function is unknown.

The preceding discussions still assume the impractical condition that sufficient design samples and classifier parameters are given. However, one should notice that GPD shows a high degree of utility and originality in practical circumstances in which only finite resources are available. In fact, it has been shown in the literature (e.g., Katagiri et al. 1998) that the more limited the design resources are, the more distinctive from others the GPD result is. A reason for this significance is the fact that GPD always directly pursues the minimum classification error situation, conditioned by given circumstances, while other traditional approaches aim at a status of the parameters that is not consistent with the minimum misclassification condition. For example, the MLE approach aims at an estimate of a sample distribution function that does not directly indicate classification results, and also, the minimization of the squared error objective, widely used in ANN design attempts, does not evaluate classification results.

The above argument is actually independent of the selection of optimization methods, such as the probabilistic descent update of GPD. Instead, an essential point included therein is the GPD formalization using the misclassification measure and the smooth classification error objective.

Links with Others In the previous subsections, we indicated that GPD design using the classification error objective is a general design method for achieving accurate recognition. In the literature (e.g., Katagiri et al. 1991a,b, 1998), the relationship between GPD training and other ANN-based training methods such as discriminative training based on the minimization of the squared error objective is clarified. In particular, it is shown that some versions of LVQ, for example, LVQ2 (Kohonen et al. 1988), can be considered as a somewhat heuristic version of the GPD implementation using the classification error objective for a multiprototype (multireference) distance network.

5.4 SPEECH RECOGNITION

5.4.1 Overview

A speech pattern is a sequence of linguistic units, such as phonemes. Accordingly, a speech recognizer should perform two operations accurately: (1) segmenting an input speech pattern to short units and (2) classifying the segmented units as preset linguistic classes. Essentially, there are obvious interactions between these two operations, and accordingly applications of ANNs are expected to contribute to both the segmentation and the classification.

In light of the discussion presented, various applications of FFNs have been investigated, some purely based on ANN concepts, and some hybrids with other system structures such as HMMs. As cited early on in this chapter, the modeling capability of FFNs is less than satisfactory for fully coping with the dynamics of speech patterns. Consequently, recognizers based only on FFNs have basically been developed to recognize short linguistic units instead of entire spoken inputs such as connected words. Two main examples of this development will be described in Section 5.4.2.

Obviously, the recognition of short units is not a goal of speech recognition. To achieve the accurate recognition of usual (long) speech inputs, state transition models such as HMMs are therefore incorporated with FFNs. Examples in this latter approach are grouped into the following two subcategories: (1) systems of integrating a highly accurate FFN-based classifier of acoustic feature vectors and an HMM-based state model, and (2) systems that are globally optimized by ANN-motivated discriminative training methods that use the squared error objective. Examples of the simple integration of an FFN-based classifier and HMM model are introduced in Section 5.4.3; Examples of global design are introduced in Section 5.4.4.

In the later part of Section 5.4, it is basically assumed that acoustic feature vectors are calculated independently from the application of ANNs.

5.4.2 Shift-Tolerant Networks

Overview As cited repeatedly, a speech signal is of a variable length and is nonstationary. Moreover, its length varies nonlinearly. To achieve the accurate classification of short linguistic units, a system should first perform the accurate segmentation of the units or a segmentation robust to its temporal shift. There are two example systems developed to cope with this difficult segmentation problem. The first is the time-delay neural network (TDNN), which has played a historic role in applications of ANNs to speech recognition in the 1980s (Waibel et al. 1989). The second is shift-tolerant LVQ (STLVQ) (McDermott and Katagiri 1991).

Time-Delay Neural Network TDNN is a special type of MLP, and its key design concept is to feed a network a stream of several successive acoustic feature vectors and constrain it to learn only from limited-sized time windows shifted over the entire speech input. After the input layer, the first intermediate layer of the network groups together features learned by the first layer and forms new, abstract features. The next intermediate layer connects this level of abstract feature representation to the output nodes, which correspond to the target phonemes of the task. An example structure of TDNN is illustrated in Fig. 5.19.

The network training is done by applying the gradient descent optimization to the squared error objective. A teaching signal that

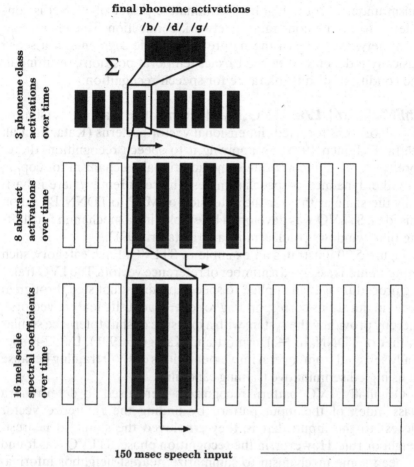

Figure 5.19 Example structure of TDNN.

indicates a correct class index in the training stage is basically of the form of a unit vector such as $(1, 0, 0)^T$ (for 3-class recognition). The squared error objective is calculated between this teaching signal vector and the output vector of TDNN. Note that the jth component of the output vector behaves as the discriminant function of C_j. After the training, TDNN classifies an input speech segment to the class having the largest discriminant function value, that is, the largest network output value.

TDNN has been compared with the standard MLE-based HMM phoneme recognizer in a Japanese phoneme recognition task and demonstrated its clear superiority (Waibel et al. 1989). A main mechanism underlying the highly discriminative power of TDNN is considered to be the constrained network connection. This enables an MLP network to learn the nature of acoustic segments in a sense basically independent of their exact temporal positions, resulting in the condition of shift tolerance for speech recognition.

Shift-Tolerant LVQ LVQ is a discriminative training version of VQ algorithms for fixed-dimensional vector patterns (Katagiri et al. 1991a; Kohonen 1995). In applying it to speech recognition, therefore, one should incorporate some additional mechanism for coping with the dynamics of speech samples. The requirement here is basically the same as that for the extension of MLP to TDNN. Based on this idea, STLVQ was developed by employing a mechanim for shifting time windows (McDermott and Katagiri 1991).

Figure 5.20 illustrates an example of STLVQ. Each category, such as phoneme, is assigned a number of reference vectors. The LVQ training procedure is then applied to speech patterns that step through in time (in durations corresponding to several acoustic feature vectors), thereby providing the system with a measure of shift tolerance similar to that of TDNN. In McDermott and Katagiri (1991), LVQ2 is specially selected from several implementations of LVQ training because of its high discriminative training capability.

Originally, LVQ pattern recognizers were used to determine a class index of the input pattern by finding the reference vector closest to the input; that is, they employed the standard nearest-neighbor rule. However, in the recognition phase, STLVQ was found to need some mechanism to summarize nearest-neighbor information; this is because it possesses multiple nearest-nighbor reference vectors, one for each time window position. STLVQ actually computes the average distance over all pairs of input and nearest-neighbor reference. Basically, this computation of the average distance behaves in a similar manner to the abstraction in TDNN.

The LVQ algorithm used in STLVQ is not an exact version of a GPD-based VQ method that has the potential of achieving the minimum classification error condition. However, based on its closeness to the minimum classification error learning, one can expect that STLVQ also shows a high discriminative power. Actually, the utility of STLVQ has successfully been demonstrated in a comparative study with TDNN (McDermott and Katagiri 1991). In the comparisons, STLVQ achieved almost the same recognition accuracy as TDNN by using a much smaller size of trainable parameters than its counterpart.

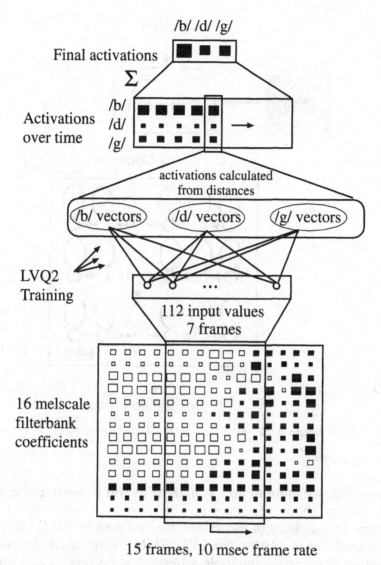

Figure 5.20 Example structure of STLVQ.

5.4.3 Hybrid Recognizers of FFN and HMM

LVQ/HMM The success in STLVQ development has naturally led to other applications of LVQ to the design of codebook vectors of discrete density HMMs (Iwamida et al. 1990; Kimber et al. 1990; Yu et al. 1990). Note that codebook vectors, associated with multinomial distributions, are basically of the same form as LVQ reference

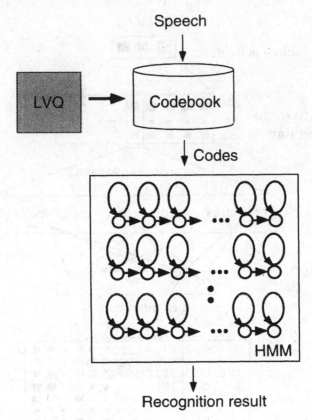

Figure 5.21 Structure of an LVQ/HMM hybrid.

vectors. The structure of an LVQ/HMM hybrid is illustrated in Fig. 5.21. The codebook vectors of this structure are trained with LVQ and the probabilities of the HMM are designed by MLE training. In Iwamida et al. (1990), the LVQ/HMM hybrid specially uses a sequence of successive multiple acoustic feature vectors as a single high-dimensional codebook vector and shows clear improvements in phoneme recognition over the conventional HMM classifier based on a traditional VQ algorithm, that is, k-means clustering.

FFN/HMM   According to the Bayes decision rule, it is known that the accurate estimation of *a posteriori* probabilities, each for deciding a class, conditioned by the observation of a sample, leads to the optimal, minimum classification error probability condition. In this light, many studies on ANNs have been conducted aiming at such accurate estimation, and it has been shown that the output of FFNs trained with the minimization of the squared error objective is a useful estimate of the *a posteriori* probability of the network input (e.g., Bourlard and Wellkens 1990; Gish 1990; Richard and Lippmann 1991).

In Bourlard and Wellkens (1990), a method of estimating the *a posteriori* probability for an observed acoustic feature vector by using an FFN system is proposed, and its utility is demonstrated in several tasks of classifying acoustic feature vectors. The system was designed using the squared error objective or the mutual information objective. The main target of the FFN system in this proposal is to classify acoustic feature vectors correctly but not to achieve the accurate classification of an entire speech input, which is usually in a word or connected word form. To utilize the discriminative power of an FFN acoustic feature vector classifier in such general speech recognition tasks, a system structure of an FFN/HMM hybrid form has been proposed (Bourlard and Wellkens 1990) [see Fig. 5.15 (right)].

The hybrid system consists of an FFN front-end module, which computes estimates of the *a posteriori* probabilities used in the acoustic feature vector classification, and an HMM module, which accumulates the probabilities to determine a class index for the entire speech input.

HMM/LVQ As cited in the previous paragraph, most FFN/HMM hybrids use the strategy of applying the discriminative power of FFNs to the acoustic feature vectors and providing the outputs of the FFNs to the post-end HMM classifiers. Another possibility of using the advantages of both the FFN and HMM is to first use the HMM to convert a variable-durational speech input to a fixed-dimensional vector pattern and then classify the resulting vector pattern by using an FFN classifier (Gao et al. 1990; Howell 1988; Katagiri and Lee 1993).

Figure 5.22 illustrates the structure of an HMM/FFN, specially developed by using LVQ in Katagiri and Lee (1993). A procedure employed here is as follows. First, a set of n-state HMMs (one HMM per word category) is used to segment acoustic feature vectors for an input speech pattern. If there are M word categories, M segmentations are generated. For each segmentation, the feature vectors assigned to each HMM state are averaged, and the averages are concatenated for each HMM, yielding M sequences of n vectors each. Then, LVQ training/classification is applied to these time-normalized vector patterns.

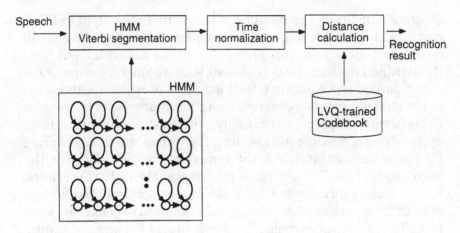

Figure 5.22 Structure of an HMM/FNN hybrid.

5.4.4 Globally Optimized Recognizers

Overview In the research trend from shift-tolerant ANNs to hybrid recognizers, the design scope was appropriately expanded and currently enables one to achieve a more realistic speech recognition system. However, it is obvious that hybrid recognizers still limit the discriminative power of ANNs only to the classification of local acoustic feature vectors instead of the classification of a whole speech input. Natural extensions of the above research scope have therefore been provided to the global optimization of hybrid systems or conventional HMM systems by using the ANN's discriminative design capability.

Among many, we focus this section on three major examples: (1) the multistate time-delay neural network (MSTDNN) (Haffner et al. 1991; Haffner 1994), (2) globally optimized FNN/HMM hybrids (Bengio et al. 1992), and (3) GPD-trained recognizers (e.g., Katagiri et al. 1998).

Multistate Time-Delay Neural Network To alleviate the design inconsistency between the TDNN module and the HMM module, both being separately designed, MSTDNN incorporates the DTW's superoperation into the training of the TDNN module. DTW is a state sequence search operation based on dynamic programming. MSTDNN applies this search operation to local TDNN phonemic or subphonemic outputs to generate an overall classification output at the word or connected word level.

The structure of MSTDNN is similar to that of other FFN/HMM hybrids (see Fig. 5.15). However, it is clearly distinctive from the others, based on the fact that MSTDNN is trained by defining an objective function for the entire speech input and consistently optimizing the whole recognizer under the criterion of reducing the objective. In contrast with the simple combinations of FFN outputs and HMM/DTW operations in the hybrids described previously, there is an interaction between the TDNN and DTW modules in the design procedure of MSTDNN. That is, in gradient descent optimization using the objective, the parameter adjustment is first propagated over the best sequence of TDNN phonemic or subphonemic output nodes and then further propagated down to the standard TDNN connections. Clearly, the resulting MSTDNN recognizer is more appropriate to the recognition of the whole speech input than the hybrids simply combining FFN classifiers and HMMs.

Globally Optimized FFN/HMM Hybrid The DTW path (state sequence) search operation is essentially the same as the Viterbi algorithm used for searching for the most probable state sequence of HMM. Accordingly, several alternatives to MSTDNN have naturally been defined by replacing DTW by the HMM and by replacing TDNN by other FFNs.

In Bengio et al. (1992), a general procedure is proposed for optimizing an ANN front-end module and an HMM post-end module consistently. The procedure can be applied to an RNN front-end module as well as an FFN front-end module; the case of the RNN/HMM hybrid is elaborated in Bengio et al. (1992). Regardless of the network-type selection, that is, an RNN or FFN, the proposed global optimization first trains the ANN front-end module as a non-linear projection function that produces an input vector sequence of the HMM module and then tunes the ANN and HMM modules under a single design objective set for accurately recognizing the whole speech input. Basically, any reasonable objective such as the squared error objective can be applied to the global optimization, though the simplest MLE-based optimization, which is not a so-called discriminative optimization, was investigated in Bengio et al. (1992).

GPD-Trained Recognizers

Segmental GPD for Continuous Speech Recognition The main subject of Section 5.4 is to introduce FFN applications to speech recognition, and no mention is made on limited GPD applications to HMM modules. However, it is worth while introducing GPD applications to HMM systems for classifying a whole speech input, spoken in a connected word form. Indeed, in such cases, a GPD-trained HMM system works as a globally optimized speech recognizer; this discovery was pursued by the development of FFN/HMM hybrids. It should be noticed, however, that the speech recognizer used in this GPD application is purely based on the standard HMM structure and also the GPD training is not applied to the feature extraction module set in front of the HMM classifier.

The formalization concept of GPD (Katagiri et al. 1991b) encompasses a wide range of application possibilities. For example, the formalization can be easily applied to the recognition of a whole speech input as well as that of acoustic feature vectors. However, each application usually needs some particular elaboration of the formalization; originally, the basis of GPD was especially elaborated for continuous (connected word) speech recognition, resulting in segmental GPD (Chou et al. 1992; Juang et al. 1997). Similar to most GPD applications, segmental GPD uses the classification error objective, which is strongly linked to the direct approximation of the minimum classification error probabilty condition.

A key issue in the segmental GPD formalization is to define a discriminant function that directly reflects the measurement computed for the recognition of the whole speech input. For an HMM recognizer, the standard HMM likelihood is used as the discriminant functions, each corresponding to an individual connected word class. The optimization procedure simply follows the original adjustment rule, introduced in Section 5.3.7. Essentially, when the recognizer misclassifies an input pattern spoken for design purposes, the HMM probabilities (and their parameters) are adjusted, aiming at the reduction of misclassification possibilities.

The effects of segmental GPD have been demonstrated in several common speech recognition tasks such as the TI digit (Chou et al. 1994). In addition, the same design concept employed in segmental GPD has been tested in several similar formalizations, demonstrating the high utility of segmental GPD (McDermott and Katagiri 1994; Rainton and Sagayama 1992).

Discriminative Feature Extraction The scope of GPD's global optimization was first embodied as discriminative feature extraction (DFE), which enables one to optimize a front-end feature extraction

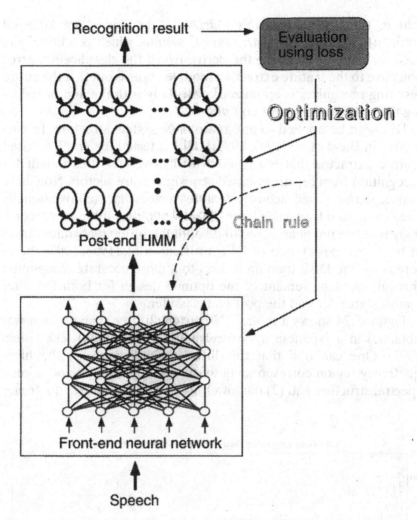

Figure 5.23 Optimization scheme of discriminative feature extraction.

module (often using a strucutre of the ANNs) and a post-end classification module (usually based on HMMs) under a single, classification error objective (e.g., Katagiri et al. 1994).

The formalization of DFE is provided by simply relaxing the GPD's adjustment scope from the classification module to the whole recognizer. Figure 5.23 illustrates a sample of DFE optimization for a hybrid recognizer consisting of an FFN-based feature extraction module and an HMM-based classification module. A point of importance here is that the trainable parameters of the feature extractor, that is, the FFN connection weights in the figure, are adjusted (trained) by using the chain rule of calculus, which is additionally used for back propagating the derivative of the classification error objective to the feature extraction module. It turns out that the entire resulting recognizer is optimized consistenly with the aim of reducing the classification error counts.

DFE can be applied to any reasonable system structure. In particular, in Biem et al. (1993, 1997), DFE is tested for an FFN-based feature extractor that is expected to discover features salient for recognition from cepstrum-based acoustic feature vectors. Note here that cepstrum-based acoustic feature vectors have conventionally been computed by using a heuristic and empirical lifter, and accordingly they are not guaranteed to be optimal for the recognition tasks at hand. An expectation of DFE in Biem et al. (1993, 1997) therefore is for the DFE training to lead to a more accurate recognition through the achievement of the optimal design for both the lifter feature extractor and the post-end classifier.

Figure 5.24 shows a typical DFE-trained lifter shape, which was obtained in a Japanese five-vowel pattern recognition task (Biem 1997). One can find that this lifter deemphasizes (1) the high-quefrency region corresponding to the pitch harmonics and minute spectral structure and (2) the lower quefrency region (0–2 quefrency

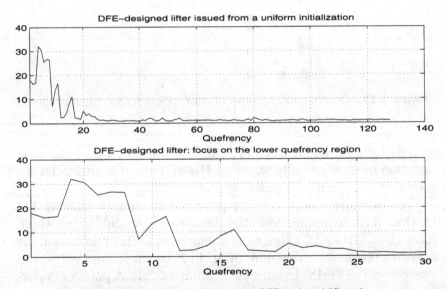

Figure 5.24 Example of a typical DFE-trained lifter shape.

region) that is dominated by the bias and slant of the overall spectrum while enhancing the 3–20 frequency region corresponding primarily to the spectral envelop structure which contains crucial information for vowel categorization. The observed shape suggests that the DFE training successfully extracted the salient features for vowel pattern recognition. In (Biem 1997), it is also reported that the training actually achieved a clear error reduction.

Focusing on the general viewpoint that feature extraction can be viewed as a process that forms a metric for measuring the class membership of an input pattern, DFE was also embodied as a discriminative metric design method (Watanabe et al. 1997a). A key point in this embodiment was to introduce different metrics, one for each class, and to optimize all of the metrics under a minimization criterion using the classification error objective.

Another embodiment of DFE has been done in the subspace method paradigm, which is one of the traditional theoretical frameworks for pattern recognition and also has some links to the present ANN formalism (Watanabe et al. 1997b). Essentially, the performance of recognizers based on the subspace method relies on the quality of the class subspaces. However, the conventional algorithms for designing subspaces, such as CLAFIC (Watanabe et al. 1967), the multiple similarity method (Iijima 1989) and the learning subspace method (Oja 1983), do not directly guarantee recognition error reduction. A DFE-based subspace method, called the minimum error learning subspace method, successfully provides a rigorous procedure for designing subspaces that guarantee the minimum recognition error condition in the sense of the GPD formalism (Watanabe and Katagiri 1997b).

GPD for Open-Vocabulary Recognition The GPD formalization concept was also embodied for open-vocabulary recognition task settings. The resulting embodiment was called minimum spotting error (MSPE) learning (Komori and Katagiri 1995). A goal of MSPE learning is to minimize spotting errors by adjusting both the keyword model λ and its corresponding threshold h. Let the trainable parameter set Λ be the combination of λ and h; $\Lambda = \{\lambda, h\}$.

For the description, let us assume λ to be an HMM keyword model. A spotting decision is fundamentally done at every time index over an input speech pattern X. Then, a discriminant function $g_t(X, S_t; \lambda)$ is defined as a function that measures the HMM probability for observing a selected speech segment S_t of input utterance X, and the spotting decision rule is formalized as: If the discriminant function meets

$$g_t(X, S_t; \lambda) > h \tag{5.22}$$

then the spotter judges at t that a keyword exists in segment S_t; no keyword is spotted otherwise.

Importantly here, this type of decision can produce, in principle, two types of spotting errors: false detections (the spotter decides that S_t does not include the keyword when S_t actually includes it) and false alarms (the spotter decides that S_t includes the keyword when S_t does not include it).

Similar to the original GPD, the MSPE formalism aims to embed the above-cited spotting decision process into an optimizable functional form and provides a concrete algorithm of optimization so that one can consequently reduce spotting errors. The spotting decision process is then emulated as *spotting measure* $d_t(X; \Lambda)$, which is defined as

$$d_t(X; \Lambda) = -h + \ln\left\{\frac{1}{|I_t|} \sum_{\varsigma \in I_t} \exp(\xi g_\varsigma(X, S_\varsigma; \lambda))\right\}^{1/\xi} \tag{5.23}$$

where I_t is a short segment, which is set for increasing the reliability and stability of the decision, around the time position t, $|I_t|$ is the size of I_t (the number of acoustic feature frames in I_t), and ζ is a positive constant. A positive value of $d_t(X; \Lambda)$ implies that at least one keyword exists in I_t and a negative value implies that no keyword exists in I_t. Decision results are evaluated by using a loss function that is defined by using two types of smoothed 0–1 functions, $\ell(\ )$ and $\hat{\ell}(\ )$, as (Komori and Katagiri 1995):

$$\ell_t(X; \Lambda) = \begin{cases} \ell'(d_t(X; \Lambda)) \\ \gamma \hat{\ell}(d_t(X; \Lambda)) \end{cases} \quad \begin{array}{l} \text{(if } I_t \text{ includes a training keyword)} \\ \text{(if } I_t \text{ includes no training keyword)} \end{array} \tag{5.24}$$

Here, $\ell_t(\ )$ approximates (1) unity for one false detection, (2) γ for one false alarm, and (3) zero (0) for a correct spotting, where γ is a parameter controlling the charateristics of the spotting decision.

The training procedure of MSPE is accordingly obtained by applying the adjustment rule (5.17) of the probabilistic descent theorem to the above-defined functions, that is, the loss, the spotting measure, and the discriminant function. In Komori and Katagiri (1995), readers can find its promising aspects, demonstrated in a task of spotting Japanese consonants.

The mechanism of keyword spotting reminds one of a hypothesis testing procedure based on Neyman–Pearson testing and the likelihood ratio test (LRT). The design concept of GPD has also been used to increase the LRT-based keyword spotting capability of an open-vocabulary speech recognizer using a filler model (Rose et al. 1995), resulting in a design procedure similar to that in Komori and Katagiri (1995). Here, the filler model is used to recognize out-of-vocabulary segments.

5.5 APPLICATIONS TO OTHER TYPES OF SPEECH PROCESSING

5.5.1 Speech Coding

Overview As explained before, speech signals are rather redundant in terms of the information needed for communications, and therefore the need for efficient methods of transmitting such information is growing dramatically in the telecommunications environment, particularly mobile communications. Figure 5.25 illustrates a basic scheme of speech coding, which consists of the following three operations:

1. Mapping an input acoustic feature vector to some codeword (index), which is *a priori* designed in a codebook
2. Transmitting the codeword from an encoding site (encoder) to a decoding site (decoder)
3. Producing an estimate of the input acoustic feature vector by using the codebook vector associated with the transmitted codeword.

Both the encoder and the decoder are assumed here to have an identical codebook, which contains (1) common codebook vectors, basically defined in the acoustic feature vector space, and (2) their corresponding codewords (indices). An input speech is first converted to a sequence of acoustic feature vectors. Each acoustic feature vector is next mapped to its closest (in the sense of the preset distortion or distance) codebook vector, and the codeword of the selected closest

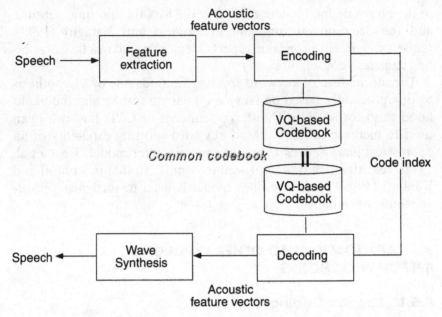

Figure 5.25 Basic scheme of speech coding.

codebook vector is transmitted to the decoder. Then, the decoder produces an estimate of the input speech by restoring the codebook vectors, each associated with the transmitted codeword.

Fundamentally, the quality of estimation depends on the configuration of the codebook, and a desirable configuration must be one that leads to minimum distortion, especially in a perceptual sense. A traditional solution to this requirement is the use of VQ algorithms. A tremendous number of studies have been conducted along this algorithmic approach, and actually VQ-based coding methods [exemplified by a code-excited linear prediction (CELP) method (Schroeder and Atal (1985)] have already formed a central part in the current telecommunications technology. However, most standard VQ algorithms are based on linear algebra, and they are not fully adequate for encoding the nonlinear dynamics of speech signals. There is therefore a distinct for ANNs to achieve a further improvement in the codebook quality by using their nonlinear transformation capability.

Among the many approaches investigated so far, we specially focus in this chapter on the following three fundamental application frameworks based on FFNs.

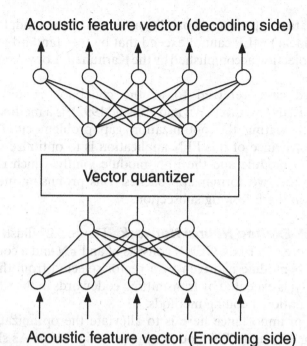

Acoustic feature vector (decoding side)

Vector quantizer

Acoustic feature vector (Encoding side)

Figure 5.26 Basic structure of speech coding using a sandglass-type MLP network.

The first case is based on the SOM network, exploiting the structural identity of SOM with VQ algorithms (e.g., Wu and Fallside 1991). One should notice that the SOM network originally consisted of a codebook. However, the FFN, that is, a type of SOM network, basically consists of linear functions; it has been revealed that the coding performance achieved by SOM learning is in the same range as that of the standard VQ algorithms.

The second case focuses on the data compression capability of the sandglass-type MLP network. The structure of this application scheme is illustrated in Fig. 5.26. The intuition here is that this type of network achieves a more efficient data compression, based on the network's nonlinearity, at a small-sized intermediate layer than standard linear VQ algorithms. However, this type of coding scheme has been shown to be less than effective due to reasons, including the following (e.g., Wu and Fallside 1992):

1. There is a gap in the design/optimization between the VQ module and the MLP module.

2. The data compression capability achieved by the sandglass-type (nonlinear) MLP cannot exceed that by the standard orthogonal projection accomplished by the Karhunen–Loeve transform.

The third case, referred to as the codebook-excited neural network (CENN) method (Wu and Fallside 1992), is a method developed for alleviating the optimization gap problem cited above. A main contrivance of this FFN application is to optimize (design) both the VQ module and the FFN module jointly, which makes it superior to the two former approaches. This promising method is elaborated in the following subsection.

Codebook-Excited Neural Network   Figure 5.27 illustrates the basic structure of CENN. CENN consists of an FFN and a codebook, and the FFN module is driven by a vector selected from the codebook. Coding is executed by transmitting codewords in the same way as the conventional coding methods.

A point of importance here is to alleviate the optimization gap between the VQ and FFN modules in the design stage. As shown in Fig. 5.27, S and $\hat{S}$ stand for the input acoustic feature vector (source

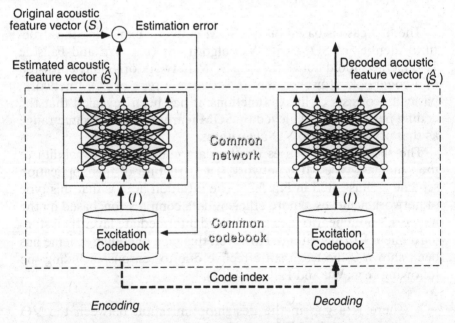

Figure 5.27   Scheme of a codebook-excited neural network.

vector) and its corresponding network output vector, respectively. In addition, let I be an excitation vector that is selected from the codebook for input S. Then, a desirable design goal becomes one of finding the status of the entire system (both the FFN module and the codebook module) that leads to the minimum distortion condition, which is measured over the training data. The design operation is executed by applying the error back-propagation algorithm to the squared error objective function, where the error is defined between source vector S and network output vector $\hat{S}$. The repetition of optimization over the training data accordingly results in a coding system, of which FFN module achieves a nonlinear transformation that is adequate for the codebook vector selection and of which the codebook pools vectors that are adequate for the associated FFN transformation.

5.5.2 Speech Enhancement

Overview Speech enhancement is a technology that aims to improve the quality or intelligibility of noisy and corrupted speech signals; its importance has been rapidly growing in recent industrial attempts to use speech technology such as speech recognition in real-world applications. Traditionally, speech enhancement was attempted by using time-domain filtering (e.g., Widrow and Stearns 1985) and spectral subtraction, which is a method of subtracting noise elements from original noisy speech signals in the acoustic feature vector domain and is currently becoming a standard speech enhancement methodology (e.g., Boll 1979). ANNs have actually been investigated to increase the performance of such existing approaches.

There are several sources of noise that need to be considered. One typical source is the additional noise that is independently added to a speech signal, for example, background noise. Basically, a speech signal corrupted by this type of noise can be modeled linearly. Another typical source of noise is the channel distortion that is usually due to low-fidelity acoustic equipment and transmission channels. In contrast to the additional noise case, the speech signal corruption due to channel distortion must be handled as a nonlinear process. Accordingly, it seems that conventional enhancement methods such as the spectral subtraction method, most of which are based on a simple linear computation mechanism, are not satisfactory for coping with the variation of noisy speech signals, in particular, the nonlinearity of the speech corruption process. Actually, even corruption by additional noise signals often becomes a nonlinear process in some transformation cases used in acoustic feature vector computation. A principal role of FFN-based applications is therefore in the addition of new nonlinear functions to existing (linear) enhancement methods.

Among several FFN-based applications investigated thus far, we introduce in this chapter the following two cases: (1) FFN-based time-domain filtering (e.g., Tamura 1987), and (2) FFN-based spectral subtraction (e.g., Sorensen 1991).

FFN-Based Filtering   FFN-based filtering in the time domain is one of the historic ANN applications to speech processing of the 1980s, and its exemplar structure employs a four-layer (two hidden layers) MLP network, as illustrated in Fig. 5.28 (Tamura 1987). The MLP-based filter in the figure is expected to accurately map a

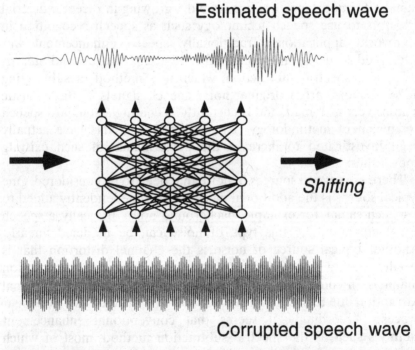

Figure 5.28   Structure of an FFN-based time-domain filtering method for speech enhancement.

windowed speech segment of the input noisy speech to an estimate of the corresponding windowed target clean speech segment. Note that the mapping is performed in the original temporal signal domain.

A goal of filtering is clearly to reduce all noise components. Therefore, network training is naturally done based on the design concept of the minimization of the squared error objective function that is defined between a network output vector (a filtered speech segment) and its corresponding teaching vector (a target clean speech segment). The minimization is accomplished by applying the back-propagation algorithm.

The main concern for FFN application is if the FFN's nonlinearity will lead to a more efficient and effective filtering than conventional methods. Indeed, in Tamura (1987), it is demonstrated that an FFN-based method achieves an interesting nonlinear feature transformation at the hidden layer and achieves a better quality of enhancement than a standard spectral subtraction approach.

FFN-Based Spectral Subtraction   As cited repeatedly, acoustic speech signals are quite redundant. Therefore, the design of time-domain filtering, which is based on the error minimization executed at every time point of signals, includes redundant and removable operations. A solution to the redundancy problem is to apply an FFN to the spectral subtraction method, which was originally implemented in a compressed feature representation framework, that is, in the acoustic feature vector space.

Figure 5.29 illustrates a basic scheme of FFN-based spectral subtraction. In this scheme, a noise-corrupted input speech signal is first converted to a sequence of acoustic feature vectors, the FFN module transforms the individual feature vectors to their corresponding new feature vectors, and finally, these new feature vectors are inversely converted to an estimate of a clean speech signal. In particular, in the case of enhancement for speech recognition, the final conversion module is replaced by a speech recognizer. An important role of the FFN is to achieve a nonlinear and highly flexible transform under the preset conditions of the acoustic feature vectors, for example, cepstrum or LPC coefficients, and the objective functions to be used for optimization. Then, the design of the network module is executed by the minimization of the objectives defined in the space of the new feature vector.

Compared to time-domain filtering using an FFN, this subtraction method is generally easier to train and more effective, mainly

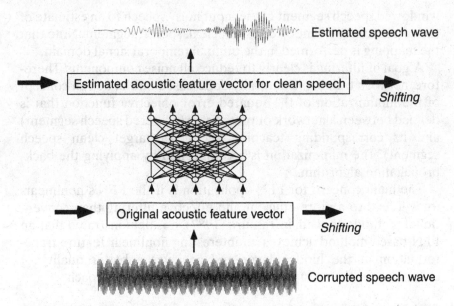

Figure 5.29 Structure of an FFN-based spectral subtraction method for speech enhancement.

because (1) it can be simply designed in the low-dimensional space of the acoustic feature vectors and their corresponding projected vectors, and (2) it can easily embed related prior knowledge, such as the perceptual significance, to the FFN-based transform function. Detailed implementations of this subtraction method and their utility are described in Sorensen (1991) and Yu et al. (1990).

Speaker Recognition

Overview   Similar to speech recognition, speaker recognition is based on the most fundamental mathematical framework of pattern recognition, that is, the Bayes' decision theory, and it is distinctively characterized by its feature selection strategy, which attempts to extract speaker identity information in place of the linguistic information that is almost always used for speech recognition.

Speaker recognition is usually grouped into two major categories: (1) speaker identification, which is the process of identifying an unknown speaker from a known population, and (2) speaker verification, which is the process of verifying the identity of a claimed speaker from a known population. Given a test utterance X, which is assumed to be represented as a sequence of acoustic feature vectors, the likelihood (possibility) of observing X being generated by speaker j is denoted as $g_j(X; \Lambda)$, where Λ is a set of trainable recognizer parameters. Then, the decision operation for speaker identification is formalized as:

$$k = \arg\max_j g_j(X; \Lambda) \qquad (5.25)$$

with k being the identified speaker, attaining the highest likelihood score among all competing speakers. The decision operation for speaker verification is formalized as: If the likelihood meets

$$g_k(X; \Lambda) > h_k \qquad (5.26)$$

then the recognizer accepts the claimed speaker identity k; the recognizer rejects it otherwise. One can easily notice here that the formalization for speaker identification is equivalent to that for basic speech recognition such as isolated word recognition and also that the formalization for speaker verification is essentially the same as that for keyword spotting, which is a basic approach to open-vocabulary speech recognition. One may also notice that the recognizer training here incurs the same difficulty in the likelihood function estimation for $g_j(X; \Lambda)$'s as does the conventional training of continuous speech recognizers and keyword spotters. Consequently, FFNs and their related technologies, such as GPD training, have been investigated, aiming at alleviating the limitations of the conventional design examples (e.g., Bennani and Gallinari 1990, 1991; Naik and Lubensky 1994; Oglesby and Mason 1991).

We introduce the following three cases: (1) neural tree recognizer, (2) FFN-based hybrid and model combination, and (3) recognizer design based on GPD.

Neural Tree Recognizer Basically, a large neural network that contains many trainable parameters has a large capability of function approximation, modeling, and discrimination. However, applying such a large network directly to a large-scale and difficult problem is not necessarily recommendable: It is often difficult to optimize large networks due to their complex objective function surfaces for difficult problems. In addition, it is a difficult question how to determine an adequate structure and size of a network for the task at hand. Therefore, several attempts based on the concept of

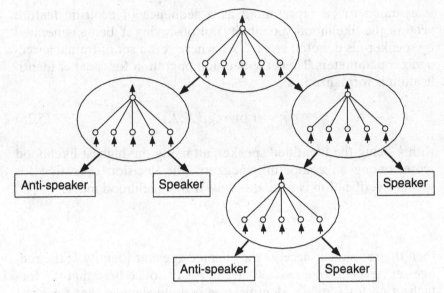

Figure 5.30 Typical structure of a neural tree network.

self-organization, which basically enables one to automatically find an appropriate network structure (and size) through training, have been investigated, aiming at alleviating these problems. One of the typical attempts is the neural tree network (NTN) (Sankar and Mammone 1993).

Figure 5.30 illustrates a typical structure of NTN. NTN is a hierarchical network classifier, and its structural concept is based on a decision tree, which is a popular decision strategy in artificial intelligence. Each node of NTN consists of a single-layer perceptron (SLP) network and the network works as a two-class (binary) classifier. Note here that the network at each node can fundamentally be a network other than SLP, but a simple SLP is preferable because of its simplicity. The NTN recognizer is applied to individual acoustic feature vectors. Given a spoken input, the input acoustic feature vectors, which are extracted from the input speech signal, are divided into two subsets at each node of every structural level. A principal assumption here is that the leaf node of NTN contains the acoustic feature vectors of a single speaker class, and therefore one can determine the speaker class of the input vectors by observing the leaf nodes or the nodes of the lower levels.

Obviously, the performance of an NTN-based speaker recognizer is determined by the behavior of the individual node networks.

Aiming at accurate recognition, the design of NTN is executed as follows. Given a set of design acoustic feature vector data at a particular node, its corresponding SLP network is trained so that it can accurately classify the data into two classes: a target speaker class and an antispeaker class. The network is then trained by applying the squared error objective to the network output and a teaching signal that indicates the true speaker class of the input acoustic feature vector, aiming at the minimum status of the objective over all of the training data. After the training, the SLP network classifies the input vector to the two classes. If all of the data within a node belong to the same speaker class, the node becomes a leaf. Otherwise, the data is further split into two subsets, which are at the adjacent low layer. Naturally, one can expect that the repetition of this splitting procedure finally achieves a complete classification of the design data. In most cases, however, the repetition is stopped based on some criterion in order to avoid the overlearning problem.

Because a binary NTN only makes a decision on whether the input acoustic feature vector belongs to the target speaker or not, a binary NTN should be prepared for each speaker. The acoustic feature vectors of a test speech signal (of an unknown speaker) are applied to all of the possible speaker NTNs. Then, a classification decision is made by measuring the size of the target speaker's leaf node, that is, the number of acoustic feature vectors fallen in the speaker leaf node. Details of implementations and experimental results are shown in Farrell et al. (1994). In Farrell et al. (1994), an alternative selection to the squared error objective, that is, the use of the classification error objective, is also investigated.

FFN-Based Hybrid and Model Combination Speaker identity information is mainly included in an average spectral envelop shape or its corresponding static representation in the acoustic feature vector domain, which is obtained by normalizing (reducing) the linguistic information. However, the information is originally spread over the entire temporal structure of the input speech signal, and therefore, in order to capture the speaker identity information automatically from the input, it is a natural and straightforward solution to apply a recognizer that is suitable for handling the dynamics of speech signals, such as an FFN/HMM hybrid recognizer.

In Naik and Lubensky (1994), an FFN/HMM hybrid speaker recognizer that consists of a three-layer MLP front-end is experimentally evaluated in a speaker recognition task using telephone speech data. In this scheme, the MLP front-end works as an estimator of the *a posteriori* probability, as in Bourlard and Wellkens (1990). In Naik and Lubensky (1994), the superiority of the hybrid system to a conventional HMM speaker recognzier is demonstrated.

The expectation underlying the development of hybrid systems is to utilize the advantages of different types of systems in a single system structure. In the pattern recognition area, another concept involving system combination has also been investigated, that is, the model ensemble and voting-based decision (e.g., Hansen and Salamon 1990). A simple statistical motivation in employing this model combination concept for pattern recognition is to decrease the instability in making classification (or recognition) decisions. Such a process may remind one of the effect of using the (ensemble) average value to represent the principal characteristics of distributed data in place of using a sample selected at random from the data. Another motivation is to utilize different types of feature representations and different types of classification decisions, each being performed by a different model or subsystem. If models used in such a combination can mutually compensate for the original limitations of the individual models, the combination can be expected to increase the discrimination power of a recognizer (e.g., Farrell et al. 1998).

Obviously, the idea of model combination essentially requires a large set of system and design resources. Probably due to this requirement, it has not been vigorously applied to speech recognition, which generally needs a large size of resources even for a single model development, but extensively investigated for speaker recognition, which can often be accomplished with a smaller set of resources than speech recognition (e.g., Farrell 1995; Schalkwyk et al. 1996; Sharma and Mammone 1996). Among many combinations, in Sharma and Mammone (1996), the combination of an NTN subsystem and a Gaussian mixture model (GMM) subsystem, which is a special type of RBF network, is evaluated in the task of verifying 30 speakers. The combination scheme used is a simple linear combination of individual subsystem outputs (estimates of *a posteriori* probabilities), and it can successfully reduce the errors of the modular subsystems to almost one third of the original error values.

Recognizer Trained by GPD Similar to most speech recognizers, speaker recongizers are often based on HMMs in place of ANNs. Accordingly, in order to increase the discrimination capability of

HMM-based recognizers, GPD has been extensively applied. For example, in Liu et al. (1995), GPD training is applied to the design of both an HMM-based speaker identification system and an HMM-based speaker verification system. The derivation of GPD-based training is fundamentally the same as that for speech recognition. It should be noticed, however, that whereas GPD is applied to the threshold training in Komori and Katagiri (1995), the threshold used for verification decisions is determined experimentally in Liu et al. (1995).

For speaker verification, Liu et al. (1995) also proposes the use of a GPD-trained normalized likelihood score. The decision rule in this case is formalized as: If the likelihood meets the condition

$$\frac{\log g_k(X;\Lambda)}{\log g_{k'}(X;\Lambda)} > h_k \tag{5.27}$$

then the recognizer accepts the claimed speaker identity k; the recognizer rejects it otherwise, where k is the claimed speaker and k' is the antispeaker (the set of speakers other than k). Clearly, the likelihood ratio of (5.27) is quite similar to the misclassification measure of GPD, and one can naturally expect that the discriminative power of the normalized score is increased by GPD training. Remarkable improvements based on GPD training are shown in Liu et al. (1995).

To see other GPD applications in speaker recognition tasks, readers are referred to recent literature such as Farrell et al. (1994), Matsui and Furui (1995), Setlur and Jacobs (1995), and Sugiyama and Kurinami (1992).

5.6 CONCLUDING REMARKS

In this chapter, we introduced feedforward network (FFN) applications to speech processing. An ANN including FFNs is principally a system structure concept, but generally it is considered to be a wider technical framework that encompasses the learning (training) strategy as well as the structural selection. Moreover, nowadays, the ANN framework is being considered to cover various, even traditional, system structures such as HMMs and prototype-based systems, while it had covered only so-called neural networks such as MLP and RNN originally. As the conceptual framework has been growing, the theoretical basis has regrettably become rather confusing.

To provide a clear technical perspective, we started this chapter by introducing fundamental issues in ANN design as well as a summary of speech signal processing technologies. In addition, we introduced GPD-related recognizer design topics, though they are not an ANN issue in the traditional narrow sense of ANN definitions. The GPD framework is quite useful, however, for designing and analyzing ANN applications to speech and speaker recognition in a more effective and efficient way.

A main application topic in this chapter was speech recognition. We only briefly introduced other topics such as speech coding and speech enhancement. A principal reason for this imbalance is because FFNs have been applied to speech/speaker recognition tasks with high vigor and actually many successful recognition results have been reported. As cited repeatedly, the speech signal is dynamic and its modeling inevitably needs a mechanism for representing its temporal structure. For speech recognition, this requirement is rather easily and naturally resolved by incorporating standard HMM and DTW systems. The high discriminative capability of FFNs is effectively utilized in this hybrid framework. In contrast to this recognition-oriented application, FFNs have not necessarily been satisfactory in other applications. In applications whose design goal is to model the acoustic nature of speech signals, such as speech coding and speech enhancement, rather than making classification decisions, the simple feedforward structure of FFNs is less than sufficient, and accordingly research interests in most ANN applications have been in the employment of RNNs in place of FFNs. It should be noted that many studies have been reported on RNN-based speech coding and enhancement technologies, though these topics are beyond the scope of this chapter (e.g., see Wan and Nelson 1997).

Notation

Special Symbols

A	set of HMM transition probabilities
a_{ij}	HMM transition probability from state i to state j
B	set of HMM output probabilities
$b_j(\mathbf{o}_t)$	HMM output probability for observing $\mathbf{o}_t$ at state j
$C(\cdot)$	classification operation
$d_k(X; \Lambda)$	misclassification measure for sample X of C_k
$d_t(X; \Lambda)$	spotting measure

$E[\cdot]$	expectation
$\mathcal{E}$	minimum classification error probability
$g_j(X; \Lambda)$	discriminant function for class C_j
$g_t(X, S_t; \lambda)$	discriminant function for keyword spotting
h	decision threshold for keyword spotting
h_k	decision threshold for speaker verification
$\mathbf{i}$	input vector to ANN node
I	output of ANN basis function
I_k	mutual information
$\lvert I_t \rvert$	size of I_t
$l_t(X; \Lambda)$	loss function for keyword spotting
$l(.)$	objective function
$L(\Lambda)$	expected loss
M	constant that expresses either the number of samples to be modeled in regression tasks or the number of possible classes in classification tasks
$\max_j p_\Lambda(C_j \vert X)$	maximum of $\{p_\Lambda(C_1 \vert X), \dots, p_\Lambda(C_M \vert X)\}$
N_S	the number of HMM states
O	output of ANN activation function
O_T	sequence of observed acoustic feature vectors (length = T)
$\mathbf{o}$	M-dimensional ANN output vector
$\mathbf{o}_t$	acoustic feature vector observed at time index t
p_Λ	probability estimate based on Λ
$\mathbf{q}$	HMM state sequence
q_t	HMM state at observation time index t
S_t, I_t	speech segment used in keyword spotting decision
$\mathbf{U}$	positive-definite matrix
$\mathbf{w}$	connection weight vector of ANN
X	input pattern to recognizers
$1(\cdot)$	indicator function
$\delta\Lambda$	adjustment amount of Λ
ε	learning weight
λ	set of HMM designable parameters, (A, B, π)
Λ	set of optimizable system parameters
Ω	entire sample space
Ω_k	partial space of Ω that causes a classification error according to the maximum *a posteriori* probability rule

π set of HMM initial state probabilities

π_i HMM initial state probability at state i

Abbreviations

ANN	artificial neural network
CENN	codebook-excited neural network
CLAFIC	class featuring information compression
DFE	discriminative feature extraction
DTW	dynamic time warping
FFN	feedforward network
GMM	Gaussian mixture model
GPD	generalized probabilistic descent
HMM	hidden Markov model
LPC	linear predictive coding
LRT	likelihood ratio test
LVQ	learning vector quantization
MLE	maximum-likelihood estimation
MLP	multilayer perceptron
MMI	maximum mutual information
MSPE	minimum spotting error
MSTDNN	multistate time-delay neural network
NTN	neural tree network
PARCOR	partial correlation
RBF	radial-basis function
RNN	recurrent neural network
SOM	self-organizing map
STLVQ	shift-tolerant learning vector quantization
TDNN	time-delay neural network
VQ	vector quantization

REFERENCES

Amari, S., 1967, "A theory of adaptive pattern classifiers," *IEEE Trans. EC*, vol. EC-16, pp. 299–307.

Bahl, L., P. Brown, P. de Souza, and R. Mercer, 1988, "A new algorithm for the estimation of hidden Markov model parameters," *IEEE Proc. ICASSP88*, vol. 1, pp. 493–496, New York.

Bengio, Y., R. De Mori, G. Flammia, and K. Kompe, 1992, "Global optimization of a neural network-hidden Markov model hybrid," *IEEE Trans. NN*, vol. 3, no. 2, pp. 252–259.

Bennani, Y., and P. Gallinari, 1990, "A connectionist approach for speaker identification," *Proc. ICASSP90*, pp. 261–268, Albuquerque, NM.

Bennani, Y., and P. Gallinari, 1991, "On the use of TDNN-extracted features information in talker identification," *Proc. ICASSP91*, pp. 385–388, Toronto, Canada.

Biem, A., S. Katagiri, and B.-H. Juang, 1993, "Discriminative feature extraction for speech recognition," IEEE, *Neural Networks for Signal Processing*, Vol. III (New York: IEEE Press), pp. 392–401.

Biem, A., S. Katagiri, and B.-H. Juang, 1997, "Pattern recognition using discriminative feature extraction," *IEEE Trans. SP*, vol. 45, pp. 500–504.

Boll, S. F., 1979, "Suppression of acoustic noise in speech using spectral subtraction," *IEEE Trans. ASSP*, Vol. ASSP-27, pp. 113–120.

Bourlard, H., and C. Wellkens, 1990, "Links between Markov models and multilayer perceptrons," *IEEE Trans. PAMI*, vol. 12, pp. 1167–1178.

Bregman, A. S., 1990, *Auditory Scene Analysis*, (Cambridge, MA: MIT Press).

Broomhead, D. S., and D. Lowe, 1998, "Multi-variable functional interpolation and adaptive networks," *Complex Syst.*, vol. 2, pp. 321–355.

Chou, W., B.-H. Juang, and C.-H. Lee, 1992, "Segmental GPD training of HMM based speech recognition," *Proc. IEEE ICASSP92*, vol. 1, pp. 473–476, San Francisco.

Chou, W., C.-H. Lee, and B.-H. Juang, 1994, "Minimum error rate training of inter-word context dependent acoustic model units in speech recognition," *Proc. ICSLP94*, pp. 439–442, Yokohama, Japan.

Denes, P. B., and E. N. Pinson, 1993, *The Speech Chain: The Physics and Biology of Spoken Language* (New York: W. H. Freeman).

Duda, R., and P. Hart, 1973, *Pattern Classification and Scene Analysis* (New York: Wiley).

Fant, G., 1973, *Speech Sounds and Features*, (Cambridge, MA: MIT Press).

Farrell, K. R., 1995, "Text-dependent speaker verification using data fusion," *Proc. ICASSP95*, pp. 349–352, Detroit.

Farrell, K. R., R. J. Mammone, and K. T. Assaleh, 1994, "Speaker recognition using neural networks and conventional classifiers," *IEEE Trans. SAP*, vol. 2, no. 1, Part II, pp. 194–205.

Farrell, K. R., R. P. Ramachandran, and R. J. Mammone, 1998, "An analysis of data fusion methods for speaker verification," *Proc. ICASSP98*, pp. 1129–1132, Seattle.

Fu, K., 1968, *Sequential Methods in Pattern Recognition and Machine Learning*, (New York: Academic Press).

Fukunaga, K., 1972, *Introduction to Statistical Pattern Recognition*, (New York: Wiley).

Funahashi, K., 1989, "On the approximate realization of continuous mappings by neural networks," *Neural Networks*, vol. 2, no. 3, pp. 183–191.

Gao, Y.-Q., T.-Y. Huang, and D.-W. Chen, 1990, "HMM-based warping in neural networks," *Proc. ICASSP90*, vol. 1, pp. 501–504, Albuquerque, NM.

Geman, S., and D. Geman, 1984, "Stochastic relaxation, Gibbs distributions and the Bayesian restoration of images," *IEEE Trans. PAMI*, vol. PAMI-6, No. 6, pp. 721–741.

Gish, H., 1990, "A probabilistic approach to the understanding and training of neural network classifiers," *IEEE, Proc. of ICASSP90*, pp. 1361–1364, Albuquerque, NM.

Gray, R., 1984, "Vector quantization," *IEEE ASSP Mag.*, vol. 1, pp. 4–29.

Haffner, P., 1994, "A new probabilistic framework for connectionist time alignment," *Proc. ICSLP94*, pp. 1559–1562, Yokohama, Japan.

Haffner, P., M. Franzini, and A. Waibel, 1991, "Integrating time alignment and neural networks for high performance continuous speech recognition," *Proc. ICASSP91*, pp. 105–108, Toronto, Canada.

Hansen, L., and P. Salamon, 1990, "Neural network ensembles," *IEEE Trans. PAMI*, vol. 12, no. 10, pp. 993–1001.

Hartman, E., J. Keeler, and J. Kowalski, 1990, "Layered neural networks with Gaussian hidden units as universal approximations," *Neural Computation*, vol. 2, pp. 210–215.

Holland, J., 1992, *Adaptation in Natural and Artificial Systems* (Cambridge, MA: MIT Press).

Howell, D., 1988, "The multi-layer perceptron as a discriminating post processor for hidden Markov networks," in *Proc. 7th FASE Symp.*, pp. 1389–1396.

Huang, X., Y. Ariki, and M. Jack, 1990, *Hidden Markov Models for Speech Recognition* (Edinburg U.K.: Edinburg Univ. Press).

Iijima, T., 1989, *Pattern Recognition Theory* (Tokyo: Morikita), in Japanese.

Iwamida, H., S. Katagiri, E. McDermott, and Y. Tohkura, 1990, "A hybrid speech recognition system using HMM's with an LVQ-trained codebook," *J. Acoust. Soc. Jpn. (E)*, vol. 11, no. 5, pp. 277–286.

Juang, B.-H., and S. Katagiri, 1992a, "Discriminative training," *ASJ, J. Acoust. Soc. Jpn. (E)*, vol. 13, no. 6, pp. 333–339.

Juang, B.-H., and S. Katagiri, 1992b, "Discriminative learning for minimum error classification," *IEEE, Trans. SP.*, vol. 40, no. 12, pp. 3043–3054.

Juang, B.-H., W. Chou, and C.-H. Lee, 1997, "Minimum classification error rate methods for speech recognition," *IEEE Trans. SAP*, vol. 5, pp. 257–265.

Katagiri, S., 1994, "A unified approach to pattern recognition," *Proc. ISANN94*, pp. 561–570, Tainan, Taiwan.

Katagiri, S., and C.-H. Lee, 1993, "A new hybrid algorithm for speech recognition based on HMM segmentation and learning vector quantization," *IEEE Trans. SAP*, vol. 1, pp. 421–430.

Katagiri, S., C.-H. Lee, and B.-H. Juang, 1991a, "Discriminative multi-layer feed-forward networks," *IEEE, Neural Networks Signal Processing*, pp. 11–20.

Katagiri, S., C.-H. Lee, and B.-H. Juang, 1991b, "New discriminative training algorithms based on the generalized probabilistic descent method," *Neural Networks for Signal Processing*, pp. 299–308.

Katagiri, S., B.-H. Juang, and A. Biem, 1994, "Discriminative feature extraction," in *Artificial Neural Networks for Speech and Vision*, R. Mammon ed. (London: Chapman and Hall), pp. 278–293.

Katagiri, S., B.-H. Juang, and C.-H. Lee, 1998, "Pattern recognition using a family of design algorithms based upon the generalized probabilistic descent method," *Proc. IEEE*, vol. 86, no. 11, pp. 2345–2373.

Kimber, D., M. Bush, and G. Tajchman, 1990, "Speaker-independent vowel classification using hidden Markov models and LVQ2," *Proc. ICASSP90*, vol. 1, pp. 487–500, Albuquerque, NM.

Kohonen, T., 1995, *Self-Organizing Feature Maps* (New York: Springer).

Kohonen, T., G. Barna, and R. Chrisley, 1988, "Statistical pattern recognition with neural networks: Benchmarking studies," *Proc. ICNN*, vol. 1, pp. I-61–I-68.

Komori, T., and S. Katagiri, 1995, "A minimum error approach to spotting-based pattern recognition," *IEICE Trans. Inform. Syst.*, Vol. E78-D, no. 8, pp. 1032–1043.

Lee, C.-H., F. K. Soong, and K. K. Paliwal, eds., 1996, *Automatic Speech and Speaker Recognition* (Norwell, MA: Kluwer).

Liu, C.-S., C.-H. Lee, W. Chou, B.-H. Juang, and A. Rosenberg, 1995, "A study on minimum error discriminative training for speaker recognition," *J. Acoustical Soc. Am.*, vol. 97, no. 1, pp. 637–648.

Matsui, T., and S. Furui, 1995, "A study of speaker adaptation based on minimum classification error training," *Proc. Eurospeech95*, pp. 81–84, Madrid, Spain.

McDermott, E., and S. Katagiri, 1991, "LVQ-based shift-tolerant phoneme recognition," *IEEE Trans. SP*, vol. 39, pp. 1398–1411.

McDermott, E., and S. Katagiri, 1994, "Prototype-based MCE/GPD training for various speech units," *Comput. Speech Language*, vol. 8, pp. 351–368.

Naik, J. M., and D. M. Lubensky, 1994, "A Hybrid HMM-MLP speaker verification algorithm for telephone speech," *Proc. ICASSP94*, pp. 153–156, Adelaide, Australia.

Nilsson, N., 1990, *The Mathematical Foundations of Learning Machine* (San Mateo, CA: Kaufmann).

Oglesby, J., and J. S. Mason, 1991, "Radial basis function networks for speaker recognition," *Proc. ICASSP91*, pp. 393–396, Toronto, Canada.

Ohyama, G., S. Katagiri, and K. Kido, 1981, "A new method of cepstrum analysis using comb type quefrency window," *J. Acoust. Soc. Jpn. (E)*, vol. 2, no. 3, pp. 135–139.

Oja, E., 1983, *Subspace Methods of Pattern Recognition* (Letchworth, England: Research Studies Press).

Park, J., and I. Sandberg, 1991, "Universal approximation using radial-basis-function networks," *Neural Computation*, vol. 3, pp. 246–257.

Rabiner, L., and B.-H. Juang, 1993, *Fundamentals of Speech Recognition* (Englewood Cliffs, NJ: Prentice-Hall).

Rainton, D., and S. Sagayama, 1992, "Minimum error classification training of HMM's—Implementation details and experimental results," *ASJ, J. Acoustical Soc. Jpn. (E)*, vol. 13, no. 6, pp. 379–387.

Richard, M. D., and R. P. Lippmann, 1991, "Neural network classifiers estimate Bayesian a posteriori probabilities," *Neural Computation*, vol. 3, no. 4, pp. 461–483.

Robinson, A. J., 1994, "An application of recurrent nets to phone probability estimation," *IEEE Trans. NN*, vol. 5, no. 2, pp. 298–305.

Rose, R. C., B.-H. Juang, and C.-H. Lee, 1995, "A training procedure for verifying string hypothesis in continuous speech recognition," *Proc. ICASSP95*, pp. 281–284, Detroit.

Rumelhart, D. E., G. E. Hinton, and R. J. Williams, 1986, "Learning internal representations by error propagation," in *Parallel Distributed Processing: Explorations in the Microstructure of Cognition*, D. E. Rumelhart et al., eds. (Cambridge, MA: MIT Press).

Sankar, A., and R. J. Mammone, 1993, "Growing and pruning neural tree networks," *IEEE Trans. Computers*, vol. C-42, pp. 221–229.

Schalkwyk, J., N. Jain, and E. Barnard, 1996, "Speaker verification with low storage requirements," *Proc. ICASSP96*, pp. 693–696, Atlanta.

Schroeder, M. R., and B. S. Atal, 1985, "Code excited linear prediction (CELP): High quality speech at low bitrates," *Proc. ICASSP85*, pp. 937–940, Tampa, FL.

Schuster, M., and K. K. Paliwal, 1997, "Bidirectional recurrent neural networks," *IEEE Trans. SP*, vol. 45, no. 11, pp. 2673–2681.

Setlur, A., and T. Jacobs, 1995, "Results of a speaker verification service trial using HMM models," *Proc. Eurospeech95*, pp. 639–642, Madrid.

Sharma, M., and R. Mammone, 1996, "Subword-based text-dependent speaker verification system with user-selectable passwords," *Proc. ICASSP96*, pp. 93–96, Atlanta, GA.

Sorenson, H., and D. Alspach, 1971, "Recursive Baysian estimation using Gaussian sums," *Automatica*, vol. 7, pp. 465–479.

Sorensen, H. B. D., 1991, "A cepstral noise reduction multi-layer neural netwrok," *Proc. ICASSP91*, pp. 933–936, Toronto.

Sugiyama, M., and K. Kurinami, 1992, "Minimal classification error optimization for a speaker mapping neural network," *IEEE, Neural Networks for Signal Processing II*, pp. 233–242.

Tamura, S., 1987, "An analysis of a noise reduction neural network," *Proc. ICASSP87*, pp. 2001–2004, Dallas.

Waibel, A., T. Hanazawa, G. Hinton, K. Shikano, and K. Lang, 1989, "Phoneme recognition using time-delay neural networks," *IEEE Trans. ASSP*, vol. 37, pp. 328–339.

Wan, E. A., and A. T. Nelson, 1997, "Neural dual extended Kalman filtering: Applications in speech enhancement and monaural blind signal separation," *IEEE, Neural Networks for Signal Processing VII*, pp. 466–475.

Watanabe, H., T. Yamaguchi, and S. Katagiri, 1997a, "Discriminative metric design for robust pattern recognition," *IEEE Trans. SP*, vol. 45, pp. 2655–2662.

Watanabe, H., and S. Katagiri, 1997b, "Subspace method for minimum error pattern recognition," *IEICE Trans. Inform. Syst.*, vol. E80 D, no. 12, pp. 1195–1204.

Watanabe, S., P. F. Lambert, C. A. Kulikowski, J. L. Buxton, and R. Walker, 1967, "Evaluation and selection of variables in pattern recognition," in *Computer and Information Sciences II*, J. T. Tou, ed. (New York: Academic Press), pp. 91–122.

Widrow, B., and S. D. Stearns, 1985, *Adaptive Signal Processing* (Englewood Cliffs, NJ: Prentice-Hall).

Wu, L., and F. Fallside, 1991, "On the design of connectionist vector quantizers," *Comp. Speech Language*, vol. 5, pp. 207–230.

Wu, L., and F. Fallside, 1992, "Source coding and vector quantization with codebook-excited neural networks," *Comp. Speech Language*, vol. 6, pp. 243–276.

Xie, F., and D. Van Compernolle, 1994, "A family of MLP based nonlinear spectral estimators for noise reduction," *Proc. ICASSP94*, vol. 2, pp. 53–56, Adelaide, Australia.

Yu, G., W. Russel, R. Schwartz, and J. Makhoul, 1990, "Discriminant continuous speech recognition," *Proc. ICASSP90*, Vol. 2, pp. 685–688, Albuquerque, NM.

INDEX